Digitizing the News

Inside Technology
edited by Wiebe E. Bijker, W. Bernard Carlson, and Trevor Pinch

A list of the series appears at the back of the book.

Digitizing the News
Innovation in Online Newspapers

Pablo J. Boczkowski

The MIT Press
Cambridge, Massachusetts
London, England

Set in New Baskerville by The MIT Press. Printed and bound in the United States of America. Printed on recycled paper.

Library of Congress Cataloging-in-Publication Data

Boczkowski, Pablo J.
Digitizing the news : innovation in online newspapers / Pablo J. Boczkowski.
p. cm. — (Inside technology)
Includes bibliographical references and index.
ISBN 0-262-02559-0 (hc : alk. paper)
1. Online newspapers. 2. Online newspapers—Technical innovations. I. Title.
II. Series.
PN4833.B63   2004
070.4—dc21

                                                                    2003056247

10 9 8 7 6 5 4 3 2 1

Para vos, por tus tangos como criaturas abandonadas. . . .

Contents

Acknowledgments

Ken Gergen convinced me a long time ago that everything meaningful grows out of relationships. This work is no exception. For starters, this inquiry would not have been possible had I not benefited from the hospitality and trust of the online newspapers that opened their doors for me to conduct fieldwork: HoustonChronicle.com, New Jersey Online, and the New York Times on the Web. Members of these organizations provided a conducive environment for social research by allowing me to observe their work practices, taking time to answer my questions, helping me in any way they could to make my job easier, and, when data collection was over, giving me feedback on texts that dealt with various aspects of this project. I would like to especially acknowledge the role played by those who most directly interacted with me in these online papers: David Galloway and Glen Golightly at HoustonChronicle.com, Carla Alford and Betsy Old at New Jersey Online, and John Haskins at the New York Times on the Web. They all went the extra mile to ensure my having a successful research experience. In addition, I would like to thank people in other media firms who also helped me in the early stages of this project, especially Rich Jaroslovsky at the Wall Street Journal Interactive Edition, Marsha Stoltman and Michael Troxler at *Editor & Publisher,* and Guillermo Culell, Julián Gallo, and Roberto Guareschi at *Diario Clarín* in Buenos Aires.

Throughout this project, I benefited from the scholarly and logistical support of three educational institutions: Cornell University, Columbia University, and the Massachusetts Institute of Technology. This study began as a doctoral dissertation in the Department of Science and Technology Studies at Cornell University, an intellectual community in which no question was too big and no theory or method was out of bounds. I spent the final three years of graduate school as a visitor in the Department of Sociology at Columbia University, which through the good offices of David Stark became a second academic home for me. My dissertation committee provided superb mentorship. Trevor Pinch, its chair, taught me how to think about the cultural life of artifacts, helped me acquire an ethnographic sensitivity, and continued to provide savvy guidance during the development of the dissertation into this book. Beneath his specific contributions lies a foundational one: Trevor believed in this project from day one, and his confidence accompanied me all these years as a powerful source of inspiration. My three other advisors went far beyond the call of duty by reading multiple drafts of chapters, meeting in cafés late in the evening after putting their kids to bed, and returning calls even in the weekends. Bruce Lewenstein enabled me to become a media scholar through his blend of intellectual sophistication and fine understanding of the practitioner's world, Ron Kline turned my curiosity about the past into a passion for history, and David Stark pushed my thinking far from established formulas by allowing me to share his own exploration of the relationships between new media and new organizational practices. Bruce, Ron, and David also generously advised me at various stages in the writing process.

Over the past two years, I have had the privilege of serving on the faculty of the Massachusetts Institute of Technology's Sloan School of Management. As a junior assistant professor, I could not have asked for more in terms of intellectual stimulation, professional mentorship, and institutional support. In addition, many of my colleagues at Sloan and other units of the Institute made direct contributions to this book. Four colleagues played decisive roles at critical junctures: Wanda Orlikowski, who helped me elaborate the logical structure of the argument; Fred Turner (now at Stanford University), who empowered me to trust my stories; John van Maanen, who urged me to write the kind of book I would like to read; and JoAnne Yates, who helped me make the narrative more fluid and accessible. Other colleagues who gave me useful feedback include Deborah Ancona, Lotte Bailyn, Paul Carlile, Joseph Dumit, Roberto Fernandez, Hugh Gusterson, Keith Hampton, Rebecca Henderson, David Kaiser, Thomas Kochan, Fiona Murray, Susan Silbey, Abha Sur, William Uricchio, and Eleanor Westney.

Others who provided feedback include Christine Borgman, Michel Callon, Jessica Cattelino, Alice Chan, Anita Chan, Noshir Contractor, Marianne de Laet, Eszter Hargittai, Geoff Fougere, Monique Girard, Bruce Gronbeck, Nick Jankowski, Sheila Jasanoff, Szabolcs Kemeny, Marion Lewenstein, Leah Lievrouw, Sonia Livingstone, Allison Macfarlane, Denis McQuail, María Victorie Murillo, Michael Schudson, and Noah Zatz. I also benefited from suggestions made by participants in meetings of the American Sociological Association, the Central States Communication Association, the European Group for Organizational Studies, the International Communication Association, and the Society for Social Studies of Science, and in seminars at Columbia University, Cornell University, the Massachusetts Institute of Technology, Northwestern University, Queen's University, the Santa Fe Institute, the Universidad de San Andrés, the University of Illinois at Urbana-Champaign, the University of Michigan at Ann Arbor, the University of Pennsylvania, the University of Texas at Austin, and York University.

I am most grateful for the helpful and constructive comments of three anonymous reviewers.

I also benefited greatly from the savvy, gracious, and efficient editorial hand of Larry Cohen at The MIT Press. Sara Meirowitz took over the project in its last stages and handled it in a most competent fashion. At two different stages in the writing process, Kay Mansfield and Paul Bethge did wonders to improve the language.

Family and friends provided much appreciated encouragement and support. They are too many to name individually, but my deepest gratitude goes to all. However, I would like to acknowledge a handful of them. Fernando Rosenberg, my alter ego, rescued me from myself whenever he could, and reminded me time and again that (to borrow from Fernando Pessoa) "navegar é preciso / viver não é preciso." Jorge Boczkowski, my brother and a scholar with a work ethic I have tried to imitate, was always present to enjoy the good times and help ease the bad ones. Aída Schwartz and Abraham Boczkowski, my parents, have nourished me with their unconditional and unfaltering love, which has anchored my sense of self and enabled me to pursue my own path.

I began working on this book a month before the birth of my daughter Sofía, and I submitted the final version to The MIT Press on her second birthday. I could not let an opportunity such as this go by without stating how much I love her. She has brought a wonderful new dimension to my life, for which I will be forever grateful.

Irina Konstantinovsky, my lover and partner for almost twenty years, contributed to this project in every imaginable way. I feel blessed for being able to share my life with her. It is to her that this book, and all of me, is dedicated.

1

Emerging Media

During the 1990s, online technologies in general and the World Wide Web in particular captured America's imagination with extraordinary intensity. This was expressed in an array of statements about major societal transformations, such as the creation of virtual communities and the coming of a new economy. In an influential book about virtual communities, Howard Rheingold argued that "whenever [computer-mediated communication] technology becomes available to people anywhere, they inevitably build virtual communities with it, just as microorganisms inevitably create colonies" (1994, p. 6). Similarly dramatic words have been uttered about the economy. "From the whirlwind of the dot com firms emerged a new economic landscape," wrote Manuel Castells (2001, p. 66). Castells added that, by resorting to the Internet "as a fundamental medium of communication and information-processing," business "adopts the network as its organizational form." "This sociotechnical transformation," he continued, "permeates throughout the entire economic system, and affects all processes of value creation, value exchange, and value distribution." (ibid.)

Discourse about the potential implications of online technologies and the World Wide Web for the mass media has also had a drastic connotation, raising the specter of radical consequences for the production and the consumption of news. Concerning news production, John Pavlik has suggested that the convergence of computers and telecommunication has brought forth a "new media system [that] embraces all forms of human communication in a digital format where the rules and constraints of the analog world no longer apply" (2001, p. xii), and that these technologies are "rapidly rewriting the traditional assumptions of newsroom organization and structure" (ibid., p. 108). Regarding news products and their consumption, Nicholas Negroponte has contended that "being digital will change the economic model of news selections, make your interest play a bigger role, and, in fact, use pieces from the cutting-room floor that did not make the cut on popular demand" (1996, p. 153). This widely debated idea of news personalization has left some scholars concerned about its potentially negative impact on civil society. For instance, in a book suggestively titled *republic.com*, Cass Sunstein has written that "a market dominated by countless versions of the 'Daily Me' would make self-government less workable [and] create a high degree of social fragmentation" (2001, p. 192).

Two themes cut across these and related reactions to what was initially called "cyberspace": (1) the predominance of accounts that concentrate on the effects of technological change and pay much less attention to the processes generating them and (2) the pervasiveness of analyses that underscore the revolutionary character of online technologies and the web and overlook the more evolutionary ways in which people often incorporate new artifacts into their lives. Paradoxically in view of its claims to novelty, this focus on revolutionary effects was also common during the early years of other major developments in mass media technology. Early witnesses of movies worried that they were going to irreversibly damage the moral character of the population by fostering both inactive use of time and primitive passions, to the point that authorities occasionally closed down theaters. The popularization of radio was also accompanied by strong claims about its "social destiny" (Douglas 1987, p. 303), including the end of demagogy, the advent of a more reflexive polity, and the rise of national unity in a country of growing diversity.

As with the case of movies, radio, and other major developments in the history of mass media technology, the focus on revolutionary effects has played a valuable role in raising our sensibility about the potentially radical consequences that online technologies and the web may have in the contemporary media landscape and in contemporary society at large. However, this focus has also been limited and limiting for at least two reasons.

First, it has made less visible that these effects derive not from how the technology's perceived properties fit anticipated social needs, but from the ways actors use it. The difference between these two modes of understanding the effects of technology becomes particularly evident when we look at the unforeseen uses of new artifacts in the history of mass media. For instance, the pioneer companies of recorded sound sold their first units as devices for recording and replaying the outcome of a common domestic activity: people playing musical instruments at home. However, in a short time, people began using phonographs to play music performed elsewhere, thus contributing to the birth of today's recording industry. The firms that did better were those that could shift focus from artifact makers to content producers.

The second limitation of the focus on revolutionary effects is that history also tells us that most of what ends up becoming unique about a new technology usually develops from how actors appropriate it from the starting point of established communication practices. The books published in the first decades after the invention of the printing press drew heavily from the content and the narrative traditions of oral storytelling, as well as from the layout and the production techniques of the hand-copied manuscript. Over time, this evolutionary appropriation of printing technology led to the construction of a communication artifact with the then-unique features of standardization and mass reproducibility—an artifact whose widespread adoption has been associated with such major transformations as the coming of the nation-state and the rise of modern science.

In this book, as an alternative to the dominant concern with technology's revolutionary effects, I look at the practices through which people working in established media appropriate technological developments that open new horizons and challenge their ways of doing things, and the products that result from this process. I pursue this alternative route not because I think the mass media's adoption of the web may not have revolutionary consequences but precisely because the potential for these consequences appears to be so significant that it is necessary to examine the often more evolutionary processes whereby they may or may not arise. I do this through a study of how American dailies have dealt with consumer-oriented[1] electronic publishing since the early 1980s, and I devote special attention to the emergence of online papers on the web in the second half of the 1990s. More precisely, I concentrate on technical, communication, and organizational practices enacted by print newspapers in their attempts to extend their delivery vehicle beyond ink on paper, such as the artifacts used to gather and disseminate information, the editorial conventions followed to tell the news, and the work processes undertaken to get the job done.

Online newspapers are a critical case of how actors situated within established media appropriate novel technical capabilities. Daily newspapers are a lucrative yet steadily declining business. At the end of the twentieth century, they exhibited profit margins higher than most industrial sectors and the largest share of advertising expenditures of all media. However, the indicators of progressive economic decline (among them

losses in penetration of the print product and share of the advertising
pie, and difficulties in attracting and retaining younger readers) have not
gone unnoticed by decision makers. These indicators have been linked
to broader socioeconomic trends that have compromised the long-term
viability of ink on paper as a delivery vehicle since the 1960s, such as rising newsprint and distribution costs, growing segmentation of consumption patterns, and the increased appeal of audiovisual media among
younger generations.

In this socioeconomic context, it is not surprising that in the early
1980s American dailies began to experiment with personal computers,
television, facsimile, and even regular telephones as alternative means of
providing information to the general public. But none of these initiatives
moved far beyond the experimental domain for more than 10 years. It
was with the popularization of the World Wide Web around 1995 that
millions of Americans began to get the news online, thus furnishing a
hospitable context for the first widely adopted nonprint newspaper. This
congruence of pressure to exploit the print business and pressure to
innovate in the nonprint domain makes online papers a decisive case of
how established media deal with new technologies.

The main thesis that results from this inquiry is synthesized in this chapter's title, "Emerging Media." It is that new media emerge by merging
existing social and material infrastructures with novel technical capabilities, a process that also unfolds in relation to broader contextual trends.
More specifically, online newspapers have emerged by merging print's
unidirectional and text-based traditions with networked computing's
interactive and (more recently) multimedia potentials. This has occurred
partly as a reaction to major socioeconomic and technological trends,
such as a changing competitive scenario and developments in computers
and telecommunications—trends that, in turn, online newspapers have
influenced. In contrast with the discourse about revolutionary effects that
has been prevalent in the dominant modes of understanding online technologies and the web, my analysis shows innovations unfolding in a more
gradual and ongoing fashion and being shaped by various combinations
of initial conditions and local contingencies.

Beyond the specifics of online newspapers, this book's main thesis
underscores the heuristic value of looking at history, locality, and process
in the emergence of a new medium. A historical perspective helps the
analyst to elicit the influence of extended longitudinal patterns in the
ways actors deal with new technologies, thus achieving a more sophisticated assessment of continuities and discontinuities in media evolution.

A focus on local dynamics invites scrutiny of the contextually contingent factors that shape actors' appropriation of novel artifacts as well as their experience of the relevant trends in the larger socioeconomic and technological milieu. An emphasis on process contributes to making more visible the ongoing practices that generate the occasionally anticipated but more often unforeseen consequences of technological change.

In one of the earliest sociological accounts of print newspapers, Robert Park wrote: "The first newspaper in America . . . was published by the postmaster. The village post office has always been a public forum, where all the affairs of the nation and the community were discussed. It was to be expected that there, in close proximity to the sources of intelligence, if anywhere, a newspaper would spring up." (1925, pp. 276–277)

The once-new technology that evolved to become an established mass medium has recently begun to appropriate the first widely adopted non-print publishing alternative in almost 300 years, and the first major new medium since the advent of television. This has triggered all sorts of speculations about upcoming transformations, such as the death of print, the replacement of newspaper companies by multimedia firms, the demise of gatekeeping, and the rise of nonlinear storytelling. However, what will ultimately spring up out of this appropriation is to us hardly as foreseeable as subsequent transformations in the postal system and the then-nascent mass medium were to readers of the first American newspaper at the dawn of the eighteenth century. What is certain, though, is that analyzing the practices that enact these transformations will help us understand how they occur, as well as the consequences they may have for the media industry and the society in which it exists.

In the remainder of this chapter, to further situate this book's argument, I look more closely at the object of inquiry, introduce the theoretical and methodological tools employed to study it, and outline the content of the chapters to come.

From Ink on Paper to Pixels on a Screen

The print newspaper is one of the oldest elements of the contemporary media landscape. According to Smith (1979), the first daily publication was *Einkommende Zeitung* [*Incoming News*], established by the bookseller Timotheus Ritzsch in Leipzig in 1650. The first issue of a print paper in what would become the United States was published 40 years later, when Benjamin Harris launched *Publick Occurrences, Both Foreign and Domestick* in Boston (Mott 1962). That was also the last issue of *Publick Occurrences*.

Because one criterion for a newspaper is periodicity, historians such as Emery and Emery (1978, p. 25) have instead called the *Boston News-Letter*, which began publishing regularly in 1704, the "first genuine American newspaper." The presence of newspapers in the United States has grown considerably since then. According to the Newspaper Association of America (2001), there were more than 1,400 daily newspapers in 2000, constituting a $59 billion industry that employed more than 440,000 people. These papers produced an aggregate weekday circulation of almost 56 million copies read by close to 55 percent of the adult population of the United States.

With dozens of millions of new copies printed every day in the United States alone, it is not surprising to find dailies almost everywhere. From living rooms to bathrooms, from offices to factories, from hospitals to hairdressers, from libraries to coffee shops, and from trains to planes, current issues of print papers are almost omnipresent inhabitants of modern life. Their ubiquity extends to familiar practices unrelated to news and advertising needs: sellers use them to wrap fish, painters to cover carpets and floors, homeless people to warm their bodies, campers to start fires, waiters and waitresses to balance unruly tables and chairs. The creation of such a ubiquitous artifact has implications not only for the information realm but also for the natural environment: it is estimated that producing the Sunday edition of the *New York Times*, for example, consumes about 27,000 trees (Baldwin, McVoy, and Steinfield 1996).

The ubiquity of newspapers is tied to their significant standardization. Despite differences in yesterday's and today's news and advertisements, two recent issues of the same paper tend to look remarkably alike. The same happens with different newspapers, to the point that visitors to a foreign country are often able to get a basic sense of the day's news by simply glancing at the local paper's headlines.[2] This standardization results from a relatively stable ensemble of technical, communication, and organizational practices.[3] Such a stable ensemble ensures that input consisting of information about often heterogeneous and unpredictable events is turned into a relatively homogeneous and predictable daily product.

This combination of age, ubiquity, and standardization endows the newspaper with a strong degree of familiarity. Perhaps none of its features is more taken for granted than the delivery vehicle, to the point of becoming part of the term used to designate the object. This is partly related to the fact that American newspapers have always told the news in ink on paper, despite experiencing significant technological change in their three centuries of existence. There have been some attempts to find alternatives to ink on paper as a delivery vehicle, some starting before the "computer revolution," such as the facsimile editions that the *Buffalo Evening News*, the *Dallas Morning News*, the *Miami Herald*, the *New York Times*, and the *St. Louis Post-Dispatch* published in the 1930s and the 1940s.[4] But the bulk of these attempts have taken place since the 1980s, in response to socioeconomic trends such as decreasing penetration, increasing costs, readers' moving to the suburbs and getting the news on the radio while driving to work, less homogenized consumer tastes' challenging mass advertising, and less interest in print products among the younger segments of the population.[5] Since then, American dailies began tinkering with options that utilized telephone, television, and/or computer technologies to communicate with their audience. However, none of these endeavors moved far beyond the experimental domain.

It was the popularization of the World Wide Web in the mid 1990s that furnished print papers with an information environment in which to create the first publishing alternative to ink on paper that achieved significant development and use. According to Abbate (1999), the Arpanet, the precursor of the Internet, became operational in 1969, and the World Wide Web was created in 1990. But their extensive social appropriation began around November 1993, when the National Center for Supercomputing Applications at the University of Illinois released Mosaic—the first graphical browser—for free download. Between 1993 and 1997, the number of web sites increased from 150 to 2.45 million (Sproull 2000), and of Internet hosts[6] from 1.3 million to almost 22 million (Chandler 2001). In the United States, by the end of the 1990s, more than 40 percent of the adult population was online (Compaine 2000b), and online advertising expenditures for 1999 reached $2.8 billion, equaling 1.3 percent of all media expenditures (Newspaper Association of America 2001).

At the time that "the web" began to become a household word in the United States, the print daily newspaper industry was quite profitable yet showing clear signs of economic decline. This decline resulted from, among other things, the trends that had propelled the industry to pursue consumer-oriented nonprint alternatives in the 1980s. On the positive side, revenues of newspaper companies grew at a 7.8 percent compounded annual rate between 1994 and 1998 (Moses 2000). In addition, the publicly traded newspaper-owning firms had a median return on revenue of 11.4 percent in 1997. This was relatively high in comparison with 3.3 percent for food, 6.1 percent for chemicals, and 9.0 percent for metal products, to name but a few large industrial sectors (Compaine 2000a). Furthermore, in 1999 the newspaper industry still had the largest share, at 20.9 percent, of advertising expenditures of all media, followed by direct mail with 18.7 percent, broadcast television with 18 percent, and radio with 12.1 percent (Newspaper Association of America 2001). On the negative side, however, newspapers' share of the advertising pie had decreased to 20.9 percent in 1999 from 29 percent in 1970 (Picard and Brody 1997). Circulation figures also looked troublesome: for instance, daily newspaper circulation per 1,000 population declined from 356 in 1950 to 305 in 1970 to 234 in 1995, which amounted to a 34 percent loss in this 45-year period (Picard and Brody 1997). Making this decline even more problematic was that readership of print newspapers was less prevalent among younger people, raising the specter that the decline might only accelerate in the coming decades as the population aged. For example, in 2000 slightly more than 40 percent of people between the ages of 18 and 34 read a newspaper daily basis, versus 53 percent of those between the ages of 35 and 44 and 66 percent of those between the ages of 55 and 64 (Newspaper Association of America 2001).

Thus, it is not surprising that many print papers launched online editions on the web during the second half of the 1990s. A handful of U.S. papers had published on the web before 1995, but this was a small number compared to the 175 that had built sites by the end of that year ("Number of papers with online edition tripled," *Editor & Publisher*, February 24, 1996, p. 39). Developments continued to move at a fast pace. A list compiled by Jackson and Paul (1998) in June 1997 included 702 U.S. dailies with web operations, almost half of the dailies in the country, and 2 years later only two of the 100 largest dailies lacked online editions on the web (Dotinga 1999). Usage of papers' sites also increased dramatically during the second half of the 1990s. For instance, the Internet traffic auditing firm Media Metrix reported that USAToday.com had 2.5 million visitors to its web site in December 1998 (Outing 1999b).[7] Three months later, Allegra Young, USAToday.com's Director of Strategy Research, stated that "on a typical weekday, the website has been averaging 923,000 unique users," and Bernard Gwertzman, editor of the New York Times on the Web, estimated that the online paper's usage was increasing by approximately 50 percent every 6 months (Outing 1999b).

Print papers' attempts to innovate on the web while still exploiting the print business provide me with a privileged window through which to examine the appropriation of novel technical capabilities by actors situated within established social and material infrastructures. Furthermore, the challenge of transforming an artifact so deeply ingrained in the everyday culture of contemporary industrialized societies brings to the fore the tensions between change and permanence that are at the heart of these appropriation practices. In addition, the combination of pre-web technical alternatives with a qualitatively different level of activity after 1995 constitutes a fruitful starting point for eliciting the dynamics of continuous and discontinuous phenomena by placing recent innovation processes within more extended patterns of change.

Theoretical and Methodological Considerations

Scholars who study technological and social change have often espoused relatively unilateral causal views, concentrating on technology's social impact or (especially in recent decades) on its social shaping. In this sense, the process of inquiry has, *a priori*, fixed either the technological or the social and turned it into an invariant explanans. However, recent work has demonstrated that material and nonmaterial elements originate, endure, and decay as a result of situated and interrelated processes of construction.[8] This kind of work seeks to "identify processes of the mutual shaping of society and technology, rather than to explain the social shaping of technology and the technical building of society" (Bijker and Bijsterveld 2000, pp. 485–486). Though this recent work has emphasized various empirical foci and conceptual dimensions, at least three common themes have been explored in studies of this type: actors' simultaneous pursuit of interdependent technological and social transformations, the ongoing character of this process, and the importance of the historical context in which it unfolds.

First, actors engaged in innovation tend to pursue interdependent technological and social transformations simultaneously. That is, they do not concentrate on either shaping the artifact or taking advantage of its social effects; they undertake both sets of actions at the same time. In one application of this insight to the study of media artifacts, Pinch (2001) has shown the extent to which the main actors involved in the construction of the analog music synthesizer simultaneously tinkered with its material elements, sound capabilities, multiple stakeholders, selling strategies, and distribution networks. Thus, because attention to the making of artifacts reveals the parallel development of their conditions for diffusion, "it is a mistake to think of a market as somehow miraculously coming into being with a new product or somehow waiting for the right

product to come along. . . . [Markets] have to be actively constructed."
(Pinch 2001, p. 392)

Second, the interweaving of technology and society is an ongoing
process. Hence, the shaping of an artifact does not stop after the emergence of a dominant design, and the conditions for the cultural consequences of its use start being created long before its initial deployment.
Moreover, in this continuous process, partial outcomes at an earlier stage
influence events at a later phase. An illustration of this matter in the case
of media can be found in a study I conducted on the making of national
identity and information infrastructures in the Argentine Mailing List, an
electronic mail distribution list of Argentines living abroad (Boczkowski
1999a). I have shown that narratives of nationhood triggered technical
transformations which then invited unexpected social changes that also
ended up destabilizing prior material arrangements. Hence, I have suggested that a mutual shaping perspective is best suited to capture the
sociomaterial dynamics of a communication environment such as the
Argentine Mailing List.

Third, cultural and material changes do not proceed in a historical vac-
uum, but are influenced by the legacy of processes that preceded them.
In other words, these changes do not occur "de novo" but are "the products of long historical processes that embed past contestations and settlements" (Reardon 2001, p. 6). Hence, the analyst has to look not only
at ongoing transformations in the artifact under study, but also at related
dynamics that happened before (sometimes long before) such an artifact
came into being. An example of this issue concerning communication
technologies is Kline's (2000) examination of how rural populations
adopted the telephone in the United States. Kline has described how
these populations used the telephone not only in some of the ways
intended by designers, but also in entirely new modes[9] such as "visiting"
on the party line. These unanticipated practices were strongly dependent
on the history and culture of rural life, and the manufacturers and telephone companies that recognized this fact altered the original designs to
accommodate users' preferences. Therefore, Kline (ibid., pp. 52–53) has
concluded that these "joint actions reinvented the telephone—both
technically and socially—as they wove it into the fabric of rural life. Farm
people used the telephone primarily to extend existing communication
practices."

In this book I view the appropriation of nonprint publishing options
by American dailies and the emergence of online newspapers as a new
medium through this lens of the mutual shaping of technological and

social change. Following Lievrouw and Livingstone (2002, p. 7), I use the word 'media'[10] to mean "information and communication technologies and their associated social contexts, incorporating: the artifacts or devices that enable and extend our abilities to communicate, the communication activities or practices we engage in to develop and use these devices, and the social arrangements or organizations that form around the devices and practices."

Media innovation unfolds through the interrelated mutations in technology, in communication, and in organization. I make sense of any of these three elements in the context of its links to the others, much like a triangle in which the function and meaning of any one side can be understood only in connection to the other two. To aid in this endeavor, I draw from conceptual resources originally developed in the fields more centrally concerned with each side of the triangle: science and technology studies, communication and media studies, and economic sociology and organization studies.[11] By locating the analytical gaze at the intersection of these usually separated fields, I show the existence of a deep ecology that links technology, communication, and organization. A new medium is what results from this ecology. Thus, understanding a new medium requires weaving a heterogeneous conceptual fabric able to illuminate the multiple elements and their complex relationships.

From this vantage point I make sense of data gathered through both ethnographic and historical methods. (See the Appendix for a more complete description of the research design.) To begin, I conducted case studies of projects undertaken by three online newsrooms aiming to exploit the web's capabilities as an information environment. I focused on these projects because, though not representative, they nonetheless expressed with great intensity the dynamics involved in appropriating novel technical capabilities from the starting point of established sociomaterial infrastructures. The projects are the New York Times on the Web's Technology section, the *Houston Chronicle*'s Virtual Voyager, and New Jersey Online's Community Connection. (New Jersey Online is a joint initiative of the *Newark Star-Ledger*, the *Trenton Times*, the *Jersey Journal*, and the television station News12 New Jersey.) The Times on the Web's Technology section aggregated all the print *Times*'s technology stories and added new content created for the online edition. Virtual Voyager produced multimedia packages of general-interest events. Community Connection was a free web publishing program for New Jersey nonprofit organizations. I spent between 4 and 5 months per case. I observed the work practices of those most directly related to the three projects under study and conducted 142 interviews with relevant actors, in addition to hundreds of informal conversations with my interviewees and others.

I also examined larger trends in the history of consumer-oriented electronic publishing initiatives by American dailies, from their computer-based efforts of the early 1980s to their use of the web in the late 1990s. To this end, I undertook archival research of the newspaper industry's trade publications from 1969 to 1999 and complemented the findings from these publications with secondary sources. Embedding ethnographic accounts within a historical sensibility helps to situate fine-grained but temporally limited case studies within more extended patterns of continuity and disruption. Furthermore, my narrative also aims to contribute to a history of media's recent evolution. Although this record-keeping function is always an important part of social inquiry, its relevance increases during the emergence of a new medium for two reasons. First, the influence of previous cultural forms and the number of paths pursued are much less visible after a new medium becomes more established. Second, the speed and scope of the technological and social changes accompanying the evolution of online newspapers have posed special challenges for the actors' own record keeping and the analysts' empirical work.

Outline of the Book

This book looks at the practices enacted by actors situated within established media to appropriate new technologies, and the new media that result from this process. I address these phenomena through a study of the attempts of American dailies to extend the delivery vehicle beyond ink on paper, with a special focus on online newspapers on the web. The overall result of this inquiry is captured in the notion that new media emerge by merging existing sociomaterial infrastructures with novel technical capabilities and in the notion that this evolution is influenced by a combination of historical conditions, local contingencies, and process dynamics. To articulate these general notions more concretely in relation to the data, I structure my account of the emergence of online newspapers in two dimensions: empirical findings about patterns of innovation shaping the different practices undertaken by the actors, and analytical insights on the construction, products, and use of media.

Two patterns receive particular attention: print newspapers' culture of innovation and online newsrooms' innovation paths.

First, American dailies have seen the recent developments in information technology through the lens of print and have tended to appropriate them under the assumption that the future would be an improved, but not radically different, version of the present. For example, they have often taken limited advantage of the multi-directional information flows afforded by networked computing, thus expanding the unidirectional mode prevalent in the industry but mostly preserving it. That is, interactivity has not been incorporated from the clean slate of a technology-driven future, but it has not been ignored either. The consequences of this particular culture of innovation have been twofold. On the one hand, print's forays beyond ink on paper have often resulted in artifacts not as innovative as those of competitors less tied to traditional media. On the other hand, the cumulative outcome has been one of tremendous change: by the end of the 1990s, online papers on the web were very different from their print counterparts.

Second, the innovation paths followed by online newsrooms trying to realize the web's interactive and multimedia capabilities have been shaped by three factors anchored in the world of print: the relationship with the print newsroom, the definition of the editorial function, and the representation of the public. Various permutations of these factors have led to different innovation paths and resulting artifacts. The endeavors that have been more successful in realizing the web's capabilities have articulated limited alignment with the print newsroom, enacted an editorial function structured around alternatives to traditional gatekeeping, and constructed their public as technically savvy information producers. In contrast, the endeavors that have ended up mostly reproducing print's modes on the web have taken place in online newsrooms that align themselves strongly with their print counterparts, structure editorial work along the lines of gatekeeping, and represent the intended end users as technically limited information consumers.

Eliciting these patterns of innovation yields three analytical insights about the construction of media, the products that result from this process, and their adoption by users. First, my inquiry suggests that the newsroom is a sociomaterial space in which artifacts matter greatly in how information is created, in who is involved in its creation, and in how the intended consumers are inscribed into the final product. To overlook the materiality of editorial work runs the risk of either missing important dynamics or misunderstanding their causes and implications. Second, because the results of newsroom practices are locally contingent, focusing exclusively on these products—the elements that constitute them, the logic governing their relationships, and the links to the larger context—and disregarding their production processes may lead analysts to misread necessity into the outcomes of contingency. Third, this study indicates that how users take up online news products is shaped by features of these products created during their production. Thus, making sense of users' online consumption of these products depends substantively upon their mostly offline construction.

To make the case for the notions that new media emerge by merging existing infrastructures and novel capabilities, and that this is best understood by emphasizing history, locality, and process, the remaining chapters present these empirical findings and analytical insights as follows.

Chapters 2 and 3 focus on how the U.S. newspaper industry dealt with consumer-oriented electronic publishing in the 1980s and the 1990s. Chapters 4–6 present three case studies of recent initiatives by online newspapers aimed at exploiting the web's interactive and multimedia potentials. Chapter 7 is devoted to drawing general conclusions and offering grounded reflections on the changing new media landscape.

Chapter 2 focuses on American dailies' attempts to go beyond ink on paper, from the early computer-based efforts to the popularization of the World Wide Web. Two major developments characterized this period. First, the 1980s was a decade of exploration: dailies tinkered with a diversity of delivery vehicles, information infrastructures, and content options, and they learned about the commercial feasibility of these endeavors by studying how users responded to them. Second, the first half of the 1990s saw a progressive narrowing of nonprint alternatives, and by 1995 American dailies had settled on the web as their consumer-oriented information environment of choice. Although newspapers continued to explore most of the other technical alternatives, the web clearly took center stage.

Chapter 3 analyzes how things evolved during the first 5 years of online newspapers on the web. This was a time of feverish activity. American dailies pursued multiple avenues in their web efforts, some merely reproducing print content on their sites, some significantly enhancing it with the addition of new information features, and some creating entirely new material using interactive and multimedia tools. The overall consequence of this multiplicity of innovation practices was a form of hedging in which newspapers diversified their bets by moving in many different directions.

The accounts presented in chapters 2 and 3 begin to elicit the ways in which American dailies have dealt with consumer-oriented electronic publishing. But, despite their value in illuminating longitudinal patterns, these accounts are less suited to shedding light on the concrete practices through which the established repertoire of print intersects with the novel horizons available in a digital networked information environment. In chapters 4–6, I examine some of these practices by presenting in-depth case studies of initiatives by online newsrooms aimed at creating content on a regular basis and taking advantage of some of the web's distinctive potentials. The analysis of these case studies concentrates on interdependent practices in three dimensions. First, I examine the communication strategies enacted in online newsrooms, concentrating mostly on gathering, processing, and delivering editorial content. Second, I consider the configuration of information architectures, focusing on media choice, interface design,[12] information and message flows, and use and development of publishing tools. Third, I discuss the coordination processes that tie together the work relationships of online newsroom personnel with their counterparts in the print newsroom, their advertising and marketing colleagues in the new media division, and their users when they co-produce content featured on the site.

Chapter 4 looks at the Technology section of the New York Times on the Web, a new daily section that aggregates all the technology stories that appear in various sections of the print paper with original content created for the web. This project began in 1996 as the online paper's effort to tinker with the novel potentials of online journalism. By the time I entered the field, more than 2 years later, it had evolved into a product that shared many of the characteristics of print journalism. The project had begun as an attempt to move beyond the translation of print into HyperText Markup Language (HTML)[13] by exploring the new territory of online journalism, but it turned into the translation of HTML into print by mostly reproducing print's ways in the creation of original content for the online environment. The chapter's oxymoronic title, "Mimetic Originality," aims to capture the tensions between permanence and change at the heart of this matter. My analysis suggests that the processes whereby the creation of newness turned into the creative production of sameness resulted from reproduction of print practices in the online newsroom, from an information architecture that reinforced continuity between print and online technologies as publishing environments, and from an articulation of alignment between the desk in the online newsroom in charge of the section and its relevant counterparts in the print newsroom.

Chapter 5 focuses on the *Houston Chronicle*'s Virtual Voyager project. Launched in April 1995, it used multimedia tools to foster vicarious experience in the form of "virtual voyages" by enabling users to be as close to the scene as possible without being there physically. The evolution of Virtual Voyager exhibited a seemingly contradictory trajectory. Successful with users and industry colleagues, it nonetheless resulted in commercial failure. These were not contradictory outcomes, but the two sides of the same innovation coin. The success with users and industry colleagues was mostly premised on tinkering with multimedia storytelling to an extent almost unparalleled during the early years of online papers on the web. This, at the same time, created a gap between the less innovative expectations and routines of the marketing and advertising staff and the sponsors they were trying to attract. The same processes that led users to be almost on the scene without actually being there also made corporate and advertiser constituencies experience multimedia journalism without fully appropriating it. More precisely, my study attributes this double sense of vicariousness to a combination of print, audiovisual, and information systems practices in editorial work, an information architecture that inscribes an exclusion of technically unsavvy users, the almost complete absence of alignment with the print newsroom, and a fluid coordination of productive activities by online newsroom personnel with the "creative" but not with the "business" groups.

Chapter 6 is devoted to Community Connection, a project undertaken by New Jersey Online to provide free web publishing services to nonprofit organizations in New Jersey. I argue that enabling users to participate directly in content production results from an alternative regime of information creation that I call "distributed construction" to signal its difference from the highly centralized mode of traditional media. My study suggests that this alternative regime involves tying together an artifact configuration that inscribes users as co-producers and enacts a multiplicity of information flows, work practices more geared to opening than controlling the gates of the site, and coordination mechanisms that support relationships of interdependence and multiple rationalities.

Chapter 7 is devoted to general conclusions. It starts by summarizing the empirical findings about patterns of innovation in online newspapers and the general analytical insights they yield into the construction, products, and adoption of new media. On the basis of these findings and insights, I conclude by offering grounded reflections on two general trends that mark the current new media landscape: the dynamics of convergence and the reconstruction of news. The proliferation of technical, communication, and organizational options in the development of online newspapers is tied to issues of media convergence, one of the most pervasive but least empirically examined tropes in new media discourse.

Most convergence rhetoric has assumed that technical changes would drive all media into a common form regulated by a single logic and has speculated about how best to characterize this product and its social implications. In contrast, my study shows that online newspapers have unfolded by merging print's old ways with the web's new potentials, in an ongoing process in which different combinations of initial conditions and local contingencies have led to divergent trajectories. This puts the argument back where it started, taking it from the "revolutionary effects" discourse associated with the convergence metaphor to the "evolutionary processes" ideas encapsulated by this chapter's title. Furthermore, online papers have been partially altering news production and products. More groups than in the typical case of print and broadcast media, from technical specialists to regular consumers, have more direct impact on the shaping of news, and this puts a premium on the coordination of tasks, goals, and resources across these groups. The content and the form of news are becoming more audience-centered, are being communicated in ongoing conversations, and are adding a micro-local focus. Thus, the news of online news is, among other things, that the news itself seems to be changing in its expansion from ink on paper to pixels on a screen.

2

Exploring and Settling: Alternatives to Print in the 1980s and the Early 1990s

This chapter focuses on how American dailies dealt with consumer-oriented electronic publishing before the popularization of the web.

The 1980s were years of exploration. Newspapers tinkered with a variety of technical and communication options, from providing directory services to personal computers to delivering news via facsimile, and learned about the commercial feasibility of these endeavors by studying how users responded to them. Several trends characterized these exploratory efforts. Dailies often reproduced features of the print artifact in the electronic environment, created little original content, tended to disregard user-generated communications, and often discontinued projects that worked well technically but failed to generate consumers' enthusiasm rapidly.

In the early 1990s, newspapers began narrowing their exploratory endeavors and, circa 1995, focused on the web as their preferred nonprint publishing environment. To make sense of the passage from multiple options to a preferred one, I utilize the word 'settling' and weave together the notions of settling a dispute, settling in, and the actors as settlers. Settling a particular dispute endows the chosen option with a certain degree of hardness that facilitates further development but acknowledges that future contingencies may lead actors to reconsider existing settlements. What contributed the most to settling the dispute in the present case was the widespread perception that the web was becoming the preferred environment for users. The more sites were built, the more the dispute seemed settled; the more settled the dispute seemed, the more sites were built. This leads to the notion of settling in as a development-oriented activity illuminating how sociotechnical options continue to unfold after the emergence of a dominant alternative, and to the notion of the actors as settlers moving into a territory new to them but having a preexisting social and material basis.

I argue that there is a common culture of innovation beneath this movement from exploring multiple options to settling on the web. That is, how newspapers appropriated the various alternatives to print expresses a culture of innovation marked by a combination of reactive, defensive, and pragmatic traits. The word 'reactive' underscores that actors followed technical and social trends rather than proactively preceding them. I use the word 'defensive' to emphasize that newspapers focused more on maintaining the territory occupied by the print franchise rather than on offensively trying to move into new areas. By 'pragmatic' I mean that they mostly sought the short-term well-being of what was identified as core business, rather more idealistically pursuing projects that seemed promising but could pay off only in the long term.

This chapter begins to tell the story of how American dailies have appropriated nonprint delivery vehicles. It shows that they neither ignored nor wholly embraced electronic publishing, but appropriated it full of contradictions. For instance, they embedded as much sameness as possible while building something supposedly new, and sometimes even wished for failure while striving for success. However, contradiction did not mean stasis. By the mid 1990s, a sizable portion of the industry had tinkered with some basic features of new media that would mark the great appeal of the web, and decision makers seemed convinced that nonprint alternatives were worth pursuing. Thus, in addition to reconstructing a segment of media's recent history, understanding how events unfolded in the period covered in this chapter is also crucial to making sense of more contemporary processes, both to locate the sources of continuity and to assess the significance of the discontinuities.

Exploring

The 1980s were exploratory years for American dailies' efforts to appropriate nonprint delivery vehicles. They pursued initiatives that involved various technologies, such as videotex, teletext, audiotex, and fax, and various kinds of content, from news to transactional material. Interested in the commercial viability of these alternatives, they kept an eye on their short-term market feasibility as well on as what these alternatives meant for their core print business.

Videotex
Of all the technical alternatives American dailies explored in this period, videotex was the option that attracted the most activity, funding,

and expectations. Developed by the British Post Office in the early 1970s,[1] videotex consisted of transmitting information stored in a computer database over telephone lines to a dedicated terminal, a television set equipped with a special decoder, or a personal computer. The various systems were developed as closed environments—the information and applications of one system could not be accessed by subscribers of others—with their interfaces presenting a series of numerically identified choices, using mostly text with a few simple graphics (figure 2.1). When the user selected one choice by means of a keyboard or a keypad, that information was sent to the computer database, which in turn transmitted the requested content back to the receiving unit.

Videotex was seen as tool and symbol of an upcoming "information society." "Just as the information society became a potential solution for the problem of fewer factories and natural resources," wrote Donald Case, "videotex would address the problem of how to implement the information society." (1994, p. 487) Not surprisingly, then, government agencies in several industrialized nations, including Canada, France, Germany, Japan, and the United Kingdom, undertook videotex efforts during the 1970s.[2] In the United States, videotex appeared on the scene later than in these other countries, the private sector taking the lead.[3]

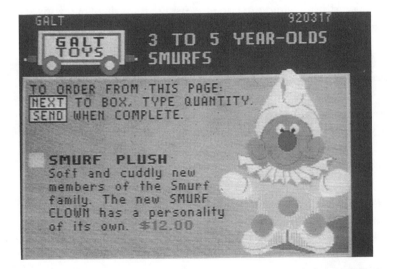

Figure 2.1
A screen of Viewtron, Knight Ridder's videotex system. Source: Alber 1985.
© Knight Ridder/Tribune Media Services. Reprinted with permission.

In this context, "newspapers were among the first to perceive the opportunities and threats posed by electronic publishing" (Baer and Greenberger 1987, p. 56). The changing socioeconomic environment, the
computerization of newspaper production, and the pervasive ideology of
technological determinism were some of the reasons for the industry's
interest in videotex.

Some of the factors mentioned in chapter 1, such as stagnant circulation, increasing newsprint costs, shifting demographics, and changing
reading and consumption habits, indicated that social and economic
transformations were challenging the viability of the daily print paper.
"The emergence of the electronic newspaper," wrote Dozier and Rice
(1984, p. 104), "draws momentum from inherent problems of pulp newspapers." In this context, electronic alternatives were seen both as a development that could worsen the situation of print papers (for instance, by
becoming a competitor in the market for classified advertisement) and as
a potential solution to the problems posed by a shifting socioeconomic
environment (for instance, by providing newspaper firms with a paperless
delivery vehicle that would be less costly to produce and distribute than
print and more appealing to television-oriented younger consumers).

The computerization of newspaper production also contributed to the
interest in videotex.[4] The extent of this process is clear: whereas in 1969
an annual survey conducted by the American Newspaper Publishers
Association found not a single newspaper using video display terminals,
in 1981 the same survey found that the 666 organizations replying to it
reported using an aggregate of more than 46,000 terminals (Weaver and
Wilhoit 1986). Marvin (1980, p. 10) commented that since such computerization had happened "behind the scenes, newspapers appear to
the public to be no different than before. This appearance is deceptive.
. . . The heart of significant technological change in the present is the
computer's transformation of print production." Although the main goal
driving this transformation was to achieve efficiencies in the production
of the print paper, this process also affected nonprint alternatives
because the editorial and advertising content that was already uploaded
to computerized databases for print production could be easily re-
deployed to videotex and other modes of electronic publishing.

Added to the mix of a changing socioeconomic environment and the
computerization of the industry was a widespread belief among decision
makers that technology would drive history.[5] For instance, at a seminar
organized in 1980 by the Newspaper Advertising Bureau to discuss alternatives for electronic publishing, Arnold Rosenfeld, editor of the *Dayton*

Daily News, said that newspapers were going to move in this direction not because their readers wanted it but to respond to a "technological imperative" (Gloede 1980, p. 30). Along similar lines, in an opinion piece titled "Newspapers: An Endangered Species," William Chilton, president and publisher of the *Charleston Gazette*, wrote: "Technology is unyielding [and] it moves in accord with its own laws and momentum, which human beings adapt to whether we want or not." (Chilton 1982, p. 31) In a keynote address to the North Carolina Library Seminar, echoing Rosenfeld's and Chilton's rhetoric, James Scofield, the chief librarian of the *St. Petersburg Times and Evening Independent*, said: "We are now in a Technological Revolution which will undoubtedly have the same impact on our world and on our lives as did the Industrial Revolution." (Scofield 1984, p. 52) This belief in a future radically altered by developments in information technologies made attention to videotex, and in some cases heavy investment in it, sensible.

In this context marked by a changing socioeconomic environment, by computerization of the industry, and by technologically deterministic rhetoric, the *Columbus Dispatch* began publishing its "Electronic Edition" on CompuServe in July 1980 (Laankaniemi 1981; Mantooth 1982). That first videotex newspaper published by an American daily was a part of a two-year test of electronic publishing coordinated by the Associated Press, a test that also included the *New York Times*, the *Washington Post*, and the *Los Angeles Times*. The mechanics of this operation were simple: once the *Dispatch* newsroom had sent the stories to the newspaper computer system for print production, editors in the electronic newsroom assigned an index category and a priority to each story, adjusted the headline to fit the CompuServe index, and sent the stories to CompuServe computers for final delivery. The content was available Monday through Saturday from 6 P.M. to 5 A.M., and all day on Sunday, at a cost of $5 per hour (*Columbus Dispatch* 1980). The test lasted until June 1982. The overall conclusion was summed up by Lawrence Blasko, director of information technology for the Associated Press: "There is no clear and present danger to newspapers from electronic delivery of information to the home."[6] Although the Associated Press did not release the full results of this experiment, from partial information that appeared in the press it is possible to list some of the reasons behind this conclusion. First, users did not access news content heavily: "Only 5 percent of the system accesses were to the news services supplied by the [Associated Press] and its member papers." (Hecht 1983, p. 61) Second, newspapers tended to offer in their online editions the same content appearing in their print editions, and this lack of original content did not satisfy users. The consensus from user feedback was that "electronic home delivery of information represents a new medium and that simply placing the 'traditional newspaper' in the system was not going to work."[7] Third, the typical "early adopter" user profile—male, young, white, upper-income, highly educated—made for a specialized audience different from the usual audience of mass-circulation publications. Fourth, there was no critical mass of personal computer users to justify large-scale investments in the short run. A cartoon that appeared in *Editor & Publisher* after the test ended suggests that perhaps at least some members of the industry received the "negative" results with some relief (figure 2.2).

Not only the larger and more resourceful papers tested the waters of electronic publishing during the early 1980s. For instance, in February 1982 the *Tiffin Advertiser-Tribune*, an Ohio daily with a circulation of less than 20,000, launched a videotex system, an enterprise that also included developing the "On Line Universal Access" software, sold in local stores, that allowed access to the service from a wide variety of personal computers (Gibson 1983). In addition to regular videotex systems, another alternative pursued by various newspaper companies was to set up public-access videotex—dedicated terminals located in places such as

Figure 2.2
A cartoon on the Associated Press–CompuServe videotex experiment. Source: *Editor & Publisher*, October 23, 1982, p. 5. Cartoon by Douglas Borgstedt.

malls, hotels, and airports. Chronicle Publishing, Harte-Hanks Communications, and Lee Enterprises had already ventured into this area by 1983.[8]

The intensity of the videotex newspaper efforts of the early 1980s was also manifested in the legislative, inter-organizational, and educational arenas. The American Newspaper Publishers Association lobbied for a bill that would exclude AT&T from providing electronic publishing information and interactive services.[9] Newspapers also participated in industry-wide groups such as the Videotex Industry Association, which in 1982 included among its 125 members the Hearst Corporation, Knight-Ridder Newspapers, and the Times Mirror Company (Radolf 1982). Institutions of higher education were also sites of videotex activities. For example, Indiana University organized a conference on electronic news delivery in 1980 (Ahlhauser 1981; "Electronic news conference set," *Editor & Publisher*, September 27, 1980, p. 45), Brigham Young University's college newspaper set up an experiment on videotex journalism in the summer of 1980 ("Test to offer in-depth news via computer," *Editor & Publisher*, July 26, 1980, p. 19), and the University of Florida established an Electronic Text Center in 1981 ("University of Florida opens electronic text center," *Editor & Publisher*, November 14, 1981, p. 38).

But no other development in this period illustrates the extent and character of videotex initiatives better than Knight-Ridder's Viewtron project.[10] After observing the evolution of the British Post Office's Prestel videotex and other related projects outside the United States, top executives at the newspaper chain became concerned about the implications that these systems could have for the American scene. James Batten, Knight-Ridder's vice-president for news, said the following before the U.S. Senate Committee on Commerce, Science and Transportation during hearings on telecommunication legislation: "In the mid 1970s, Knight-Ridder began a careful exploration of the various new electronic information technologies sprouting on the horizon. Our initial concern was defensive. For years there's been talk about 'the electronic newspaper.' We were concerned that some of these new systems might someday represent a competitive threat to the daily newspaper. And if that was to be the case, we wanted to discover that fact earlier rather than later." (Batten 1981, p. 18)

Knight-Ridder established a wholly owned subsidiary, Viewdata Corporation of America, which in 1979 announced a joint venture with AT&T to develop a videotex system modeled after Prestel ("K-R plans

1980 pilot test for Viewtron," *Editor & Publisher*, April 21, 1979, p. 117).
AT&T provided the hardware, including the dedicated terminals used by
consumers; Knight-Ridder was in charge of the rest, including content,
advertising, marketing, and general management. In July 1980 the company began a field test in which several dozen families of Coral Gables,
an upscale Miami neighborhood, were given free access to Viewtron several weeks at a time ("Viewtron test started by Knight-Ridder," *Editor &
Publisher*, July 26, 1980, pp. 18–19). The system consisted of approximately 15,000 screens of news, information, and services, such as shopping and banking, transmitted over conventional telephone lines to a
specially modified TV set equipped with a keyboard.[11] Consumers used
the keyboard to select from various choices presented through an interface composed of text and some graphics (see figure 2.1).[12] The notion
of an active user, as opposed to the more passive one of traditional print
and broadcast media, was seen by James Batten as a distinctive feature of
Viewtron: "The system puts the viewer in the driver's seat. He asks for
what he wants, when he wants it." (Radolf 1980a, p. 8)

Knight-Ridder expressed great enthusiasm during the field test. At a
gathering of more than 100 financial analysts in the Downtown Athletic
Club in Manhattan, Viewdata Corporation of America's president Al
Gilen said: "Our research is encouraging. . . . People use Viewtron for
news, shopping, and sending messages" and announced plans for a full
market trial after the field test ended (Radolf 1980b, p. 23). The field test
ended in October 1981. Knight-Ridder was satisfied with the interest in
news and with the usage of shopping and banking capabilities, as well as
with consumers' expressions of interest in the technology and willingness
to pay for the service (Silverstein 1983).

When the test formally ended, plans for a market trial in the Miami
area, where Knight-Ridder had its headquarters, were already under way.
As opposed to the field test, this second phase would have consumers
paying for the reception unit and the use of the system and would have
advertisers paying for promotional messages. This change of focus was
common in videotex developments around that time. "Unlike the tests
that started in 1978, the current wave of experimentation is designed to
answer crucial marketing rather than technical questions," wrote Jeffrey
Silverstein (1983, p. 57). The logistics of this second phase included
opening up advertising sales offices in Chicago, Miami, and New York
and hiring a national advertisement sales force. Furthermore, between
1981 and 1983 Knight-Ridder sold Viewtron franchises to newspaper
companies around the country; partners were given access to market trial

data and the rights to launch their own local Viewtron system (Compaine 1984). This phase also involved changes in the reception unit. In July 1983, AT&T unveiled its Sceptre equipment. Used to add videotex capabilities to the television set, it consisted of a wireless keypad with a Qwerty keyboard, a communication link with a speed of 1,200 baud, and the option of connection to a printer ("American Bell unveils videotex terminal," *Editor & Publisher*, July 2, 1983, p. 29).[13]

The market trial began in October 1983 ("Knight-Ridder's Viewtron available to public this week," *Editor & Publisher*, October 29, 1983, p. 16). Six months later there were only 1,500 subscribers. Knight-Ridder had expected 5,000 subscribers in the first year, so access rates were lowered ("Viewtron growth has been slow," *Editor & Publisher*, May 19, 1984, p. 38). At the end of the first year, there were 2,800 subscribers, so Knight-Ridder laid off 41 full-time employees and announced plans to make Viewtron available through personal computers ("Knight-Ridder cuts Viewtron staff," *Editor & Publisher*, November 3, 1984, p. 32).[14] This last move was a reaction to the rise of personal computer technologies, a process that evolved so rapidly that took even computer industry insiders by surprise. Speaking about IBM's decision to enter the personal computer business, Alfred Chandler Jr. noted: "No one anticipated just how explosive that demand would be. . . . More systems were shipped over the first five months of 1983 than in all of 1982. . . . In 1984 soaring revenues had reached an estimated $5 billion, equivalent to those of the seventy-fifth largest company on the *Fortune* 500 list." (2001, p. 138)

As a result of this growth, personal computers not only penetrated the workplace but also began to become important elements in the information infrastructure of homes, a critical site for the expected consumption of videotex newspapers. According to Lee Sproull (2000), 8 percent of American households had a personal computer by 1984, a year that marked a change in the meaning of home computing. "In the first era of home computing (1977–1984), stand-alone machines were acquired for entertainment and self-improvement. In the second era (1984–1994), household machines began to be connected to online databases and to people in distant areas." (ibid., p. 260)

Although it took place at the beginning of the second era of home computing, Viewtron's move to extend its service to personal computer users came too late to stop the downhill trend.[15] Knight-Ridder announced the end of Viewtron in March 1986 ("Knight-Ridder 'pulls plug' on its videotex operation, *Editor & Publisher*, March 29, 1986, p. 14). It had grown to 20,000 users nationally,[16] but Viewtron was losing great sums of money; although no official figures were made available, Knight-Ridder reportedly lost approximately $50 million in this venture.[17] James Batten, then president of Knight-Ridder, put an end to Viewtron using words that echoed those uttered at the end of the Associated Press-CompuServe test: "Videotex is not likely to be a threat to newspapers in the foreseeable future" (ibid.)

Although Knight-Ridder did not reveal much of the market trial's data, some information moved into the public domain over the years, helping to account for Viewtron's demise. First, the service was rather slow, to the point that occasionally company officials would find copies of Viewtron ads with the slogan "The waiting is over" crossed out by users and turned into "The waiting has just begun" (Fidler 1997). The slow performance, coupled with the high cost of the access and reception unit, reduced the service's appeal to users who had to pay for it rather than receiving it for free during the 1980–81 field test. Moreover, it was difficult to retain usage after the first weeks; people subscribed, began using the system, but ceased doing so when the novelty wore off (Ashe 1991). Finally, the system did not appropriately serve users' demand for interpersonal communication.[18] "In retrospect," Roger Fidler, a member of the Viewtron team, reflected, "the interviews and usage data clearly revealed that access to databases of general news, information, and advertising was less exciting to subscribers than the ability to easily communicate with other subscribers. But that was not what anyone was prepared to hear at this time. Nearly everyone involved in the trial saw Viewtron as an advertiser-supported electronic newspaper. Its potential role as an interpersonal communication medium was considered secondary." (1997, p. 148)

By the time of Viewtron demise, there was less enthusiasm among the people engaged with videotex. For instance, at the "Videotex '84" conference there were several manifestations of this change. A reporter covering the event for *Editor & Publisher* summarized the collective mood as follows: "Videotex these days is like the grand piano on the porch—it's nice, but how do you get it into the house?" (Fitzgerald 1984, p. 36). David Simons, president of Video Corporation, stated at one of the conference's seminars that Videotex was "like a religion" in that "it requires faith and is nonprofit" (ibid., p. 36). A year later, Jodi Greenblatt, director of electronic publishing for Aspen Systems Corporation, used a telling metaphor: "Traditional videotex up to now is an abortion. Never born. Never died. Just conceived." (Miller 1985b, p. 26) In this climate it is not surprising that newspaper-related videotex efforts decreased considerably during the second half of the 1980s. However, they did not disappear entirely. A handful of modest bulletin-board services that used the personal computer as the reception unit persisted in this period. For instance, in 1982 the *Fort Worth Star Telegram* launched StarText (a text-only, ASCII-based videotex system) in partnership with the Tandy Corporation (which left the enterprise a year later). StarText provided news, classified ads, electronic banking, and electronic mail, among other services, to personal computer users for a relatively low flat fee. By 1985 it had more than 1,500 users and was growing at an annual rate of 60 percent (Miller 1985b). Three years later the system had more than 3,200 users, provided an expanded array of information through a still text-only interface, and had a 30 percent cash flow return on investment; it had became profitable after reaching 2,000 users (Miller 1988). Despite this and a few similar initiatives, interest in videotex among newspaper companies remained minimal for the rest of the decade.

Other Technical Alternatives: Teletext, Audiotex, and Fax

During the 1980s, American dailies' exploration of alternatives to print were not limited to videotex; they also included other information environments, such as teletext, audiotex, and fax papers. Teletext, another technology developed in the United Kingdom (McIntyre 1983), was a major technical option among newspapers in the first part of the 1980s. It consisted of transmitting text and rudimentary graphics over television broadcast signals,[19] taking advantage of the then-unused portion of the signal known as the "vertical blanking interval."[20] The information was first digitally encoded and stored in a computer as was done for a story to appear in print or videotex outlets, and each unit of information was indexed and identified with a number or code. Then, the content was multiplexed into the television signal at the precise point of the vertical blanking interval and was transmitted cyclically at regular intervals. Viewers who possessed a television set and a special decoder equipped with a numerical keypad could access the information transmitted by switching off the regular television show. The viewer was offered a menu of the content available in a particular teletext program—each piece of news being identified with its number—and requested the items of interest by entering the appropriate code in the keypad (figure 2.3). Because the content of the program was sent cyclically and at regular intervals, when the item requested by the viewer was being transmitted the decoder grabbed it and displayed it on the viewer's screen. Thus, teletext differed from videotex in that in the former information was transmitted over

Figure 2.3
Menu, decoder, and keypad for Ceefax, BBC's teletext system. Source: McIntyre
1980, p. 34. © British Broadcasting Corporation. Reprinted with permission.

broadcast signals and was communicated in one-way fashion—the same
content was sent to every reception unit, and each person saw only the
items he or she had requested using the keypad.

In 1980 several American dailies got involved with teletext. The
Washington Post and the *Washington Star* provided content to a teletext
field test available in the Public Broadcasting System's WETA-TV ("New
teletext system test this fall," *Editor & Publisher*, June 28, 1980, p. 18).[21]
The same year, the *Jacksonville Times-Union* and the *Jacksonville Journal* provided the local Cablevision 12 channel with a teletext magazine that contained as many as 128 screens of information, changing at 12-second
intervals ("Jacksonville dailies supply information for local cable," *Editor
& Publisher*, May 24, 1980, p. 40). Newspapers that followed the same
path in 1981 and 1982 included the *Danbury New-Times*, the *Louisville
Courier*, the *Louisville Journal*, the *Worcester Telegram*, the *Worcester Evening
Gazette*, the *Milwaukee Journal*, and the *Milwaukee Sentinel* (Abbott 1982;
Huenergard 1982; "Ottaway's Cable News to add teletext," *Editor &
Publisher*, April 11, 1981, p. 39).

But perhaps no other paper was more ambitious in its teletext endeavors than the *Chicago Sun-Times*. In April 1981, Field Enterprises Inc., publisher of the *Sun-Times*, created a new subsidiary, Field Electronic
Publishing, which then began a one-year test of Keyfax in association with
the Chicago television station WFLD (Huenergard 1981). Keyfax was pro-

duced by a dedicated newsroom staff of more than a dozen people. Among other things, the service featured the magazine "Nite Owl," which was available to WFLD-TV viewers at no additional charge and without the need of a decoder every morning from midnight to 6 o'clock. "Nite Owl" carried approximately 60 screens or frames containing updated news, weather, and ads. By 1982 the magazine was reportedly attracting between 35,000 and 75,000 viewers every night in the Chicago area (Silverstein 1983). Encouraged by this success, Field Electronic Publishing, in association with the phone company Centel Corporation and the computer manufacturer Honeywell Incorporated, attempted to deliver Keyfax nationally to the 20 million subscribers of WTBS, Ted Turner's cable station, for an additional fee. As happened with other teletext initiatives, users were unwilling to pay, and by late 1984 Keyfax was terminated.

The lack of commercial success of teletext and videotex systems reassured newspapers that ink on paper was not going to be under attack in the short term. Whereas videotex experienced a resurgence in the 1990s, the mid 1980s brought the end of large-scale teletext experiments at the same time that a few newspapers began exploring a more widespread delivery vehicle: the telephone. Audiotex (also known as "voice information systems") provided automated content over the telephone to callers looking for information in categories ranging from breaking news to weather to classified ads. Newspapers published phone numbers corresponding to the categories of content in their print editions. These services made money either by having the caller listen to an advertising message or by charging the caller. American newspapers began tinkering with audiotex around the mid 1980s. A 1985 report of the American Newspaper Publishers Association showed 19 newspaper companies involved in audiotex projects (Miller 1985a).

By 1990, approximately 50 newspapers were offering audiotex services, according to a survey by the Audiotex Group (a consulting firm), and the VRU Group (a vendor of audiotex equipment) (Fitzgerald 1990b). Usage was also on the rise; for instance, *USA Today*'s Sports Hotline received 2.3 million calls in 1988, and the *Atlanta Journal and Constitution*'s audiotex service received 9 million calls in 1990—an equivalent of 17 calls per newspaper subscriber, up from 5.2 million calls the year before ("Audiotex system joins Atlanta newspapers," *Editor & Publisher*, February 17, 1990 p. 29; Garneau 1989a; Smith 1991). Despite this growth and despite the fact that some audiotex ventures yielded modest profits, audiotex failed to generate the same level of enthusiasm that videotex, or even teletext, had conveyed in the early 1980s. Although it was not a very risky proposition financially, the payoff was not huge either, and, in general, communicating information on the phone did not seem likely to evolve into anything more than an auxiliary service. Talking about the *Atlanta Journal and Constitution*'s audiotex services, Chris Jennewin, the paper's general manager of voice information systems, commented: "These services are not particularly profitable, but they serve the readers. . . . It's a lot like color. You don't make a profit using color but it helps maintain readers." (Smith 1991, p. 2TC)

After failing commercially in the 1930s and the 1940s (Shefrin 1949; Hotaling 1948), fax papers were reborn in the late 1980s, coinciding somewhat with the rise in penetration of fax machines into the workplace. Fax papers were usually targeted to specific audiences, such as business people interested in obtaining the latest information as soon as the markets closed and travelers, executives, and diplomats eager for home news while abroad. For instance, in April 1989 the *Hartford Courant* began sending its one-page FaxPaper to readers willing to pay $2,500 a year to receive the latest news every afternoon (Radolf 1989). Not all fax papers intended for business people were so expensive: Tribfax, a product of the *Chicago Tribune* was launched in April 1990 for $400 per year (*"Chicago Tribune* introduces its fax newspaper," *Editor & Publisher*, March 31, 1990, p. 56). The *Minneapolis Star Tribune* also published a fax edition (ExecutiveFax) aimed at business people (Rosenberg 1990). Among the fax papers for travelers, perhaps the most popular was the *New York Times*'s TimesFax, launched in 1990. Consisting of four to six pages summarizing the paper's news, features, and editorials of each day, it was aimed at reaching people who could not receive the paper on a daily basis (*"N.Y. Times* starts fax service to the Far East," *Editor & Publisher*, January 27, 1990, p. 18). Despite these and other efforts throughout the first half of the 1990s, fax papers never took off. Several of them folded for lack of profits, including Tribfax and ExecutiveFax ("Fax bulletin discontinued," *Editor & Publisher*, January 12, 1991, p. 39; Fitzgerald 1990a), and most of those that continued publishing became neither significant revenue centers nor objects of much enthusiasm among those interested in a digital future.

The Technical, Editorial, and Commercial Dimensions of Exploring Alternatives to Print

The 1980s was a decade of exploration for American dailies wishing to extend beyond ink on paper. They explored technical, content, and commercial issues. First, they tried an array of delivery vehicles: stand-alone visual home devices, modified television sets, personal computers, public-access videotex units, fax machines, and conventional telephones. CD-ROMs and portable digital assistants were added a few years later. Although personal computers became the dominant alternative by the mid 1990s for reasons that seem quite logical from today's standpoint, this knowledge should not be used to read history backwards: none of these delivery vehicles seemed an obvious choice for the actors struggling to make sense of an utterly complex and uncertain situation. Otherwise there would not have been as much variety as there was. Even if some options, such as audiotex and fax newspapers, did not generate the same level of enthusiasm as videotex or teletext, that they were serious electronic choices for dozens of newspaper companies during the second half of the 1980s is an indication that the delivery vehicle was an open question for the actors throughout that decade. In the initiatives described above, newspapers tended to project features of the print paper in the electronic environment. That is, they usually replicated existing information artifacts and practices, rather than creating something different. Thus, according to Blomquist (1985, p. 424), newspapers, and other established corporations, were "more likely to cultivate conservative uses of new technology that are variations on things they already know rather than radically departing from traditional media fare." For instance, the larger videotex and teletext projects—Gateway, Keyfax, and Viewtron—began by developing a technical context based on two elements: a dedicated terminal and a closed information environment. That technical context allowed for the kind of content centralization and control that papers enjoy in print. Consumers of one system could not use their terminals to do anything else but access that system and were not able to browse the content and applications available in any of the other systems. "Videotex," Carveth, Owers, and Alexander noted, "was a top-down model of centralized providers furnishing centralized services through a specialized terminal. The design reflected the technology of the time, but it also then emphasized centralized (i.e., transaction- and information-oriented) interactive services." (1998, p. 251)

The second dimension of these exploratory activities had to do with content. Newspapers included a variety of content in their electronic editions. However, at least two significant trends emerged in the 1980s: very limited original content and low appropriation of user-authored material. Concerning the news, which is print papers' core editorial product, research has shown that videotex and teletext did not carry much original reporting, let alone any that would try to exploit the distinctive potentials of the electronic platform, but that editorial work centered on rewriting copy coming from the information providers. For instance, in her dissertation study of the Associated Press and CompuServe videotex experiment, Sara Mantooth (1982, p. 90) found that "the news stories were for the most part word-for-word duplicates, and aside from the headlines which had to be rewritten from the print version, the only really different face of the electronic newspapers was . . . the long-term storage of such items as reviews and recipes." Moreover, Brown and Atwater (1986, p. 558) concluded that in the cases of Gateway, Keyfax, and Viewtron "wire services and newspapers were the sources of news stories on all three services" and that "no evidence of stories originating from the videotex staffs was found." This was also consistent with Weaver's conclusions about the European scene: "Journalists working for teletext systems in the United Kingdom and the Netherlands do almost no independent reporting. They rely primarily upon what others are writing in various wire services and newspapers." (1983, p. 53)

Most newspapers had difficulty appropriating content generated by users. Indeed, many newspapers either did not provide technical means for this to happen, and those that did neglected its significance. This was all the more important insofar as both producing and consuming this kind of content were highly popular among videotex users. "In the early trials," Aumente (1987, p. 57) noted, "Viewtron was cool to the idea of even including an alphanumeric keyboard, the heart of such interactivity, but AT&T prevailed. The messages might be amateurish but they fostered self-expression and interactivity and were major drawing cards. But for videotex systems run by corporations steeped in traditional newspaper publishing where a dozen letters-to-the-editor daily are a lot, such forays into interactive expression must have seemed alien."

Newspapers also explored a commercial dimension of electronic publishing during this period. Although the specific foci and results of these commercial explorations were usually considered proprietary information and kept from the public eye, over the years it became known that learning about their users' interests and practices was a key motivation for newspapers venturing into the electronic domain. Red Burns of New York University's Alternate Media Center went so far as to say that the teletext project the center ran on Washington's WETA public television station was a "user trial, not a technology trial" (Mecca 1981, p. 20). From the actors' standpoint, perhaps the main finding about users emerging from these explorations was that the majority of them did not seem willing to pay for videotex, teletext, or fax papers, and, as we saw, audiotex was not a significant source of revenue. This lack of commercial success was, in turn, the decisive factor preventing the continued evolution and expansion of many videotex, teletext, audiotex, and fax initiatives during the 1980s. Thus, in general, the projects that were terminated worked well from the technical standpoint but were not profitable and, for the actors, had no clear chance of becoming so in the short run.

Settling

The first part of the 1990s was a period of renewed enthusiasm in electronic publishing, albeit from a somewhat different perspective than that of the 1980s. Although print papers kept exploring diverse alternatives, they progressively focused their resources on specific options. Despite concurrent developments in audiotex, CD-ROM, fax, online services, and portable digital assistants, the second half of the decade witnessed the provision of information and applications on the web as the preferred choice of American dailies for extending their franchise beyond ink on paper.

Narrowing Down

The early 1990s saw a slow but steady increase in videotex activities. The growth in penetration of personal computers into workplaces and homes contributed to this trend. For instance, sales of personal computers worldwide increased from about $10 billion in 1985 to about $40 billion in 1990 (McKenney 1995), and their household penetration in the United States moved from 8 percent in 1984 to 24 percent in 1993 (Sproull 2000). In addition, the parallel growth in users of online services, such as Prodigy and America Online, and a broadening of the services they offered also created a fertile soil for online newspapers. For instance, online services had 2 million subscribers in the United States in 1989, a figure that jumped to 3.8 million in 1994 and amounted to 17 percent of households with personal computers (Sproull 2000). In the case of newspapers, this evolution became apparent in the launch of the Tribune Corporation's "Chicago Online" and the *San Jose Mercury News*'s "Mercury Center," both on America Online, in 1992 and 1993 respectively (Conniff 1992; Stein 1993).

Another development that influenced this trend was the lifting of the ban that prohibited telecommunication companies from entering the electronic publishing market (Garneau 1989b). The move of the "baby Bells" into that market stimulated videotex endeavors in two ways. First, several regional Bell operating companies moved to establish what became known as "gateway" online services, which basically acted as infrastructure warehouses that minimized the investment needed for a newspaper to start a videotex system. Second, fear of competition in the provision of content by the baby Bells triggered more involvement from newspapers. In an *Editor & Publisher* article summarizing the year 1989 for the newspaper industry, we read this: "'Plastics,' whispered the family friend to Dustin Hoffman in the [1967] movie *The Graduate.* 'Telecommunications' shouted the newspaper industry last year." Therefore "[its] new prime directive became: keeping the regional Bell operating companies out of electronic publishing and testing the waters itself" (Garneau 1990, p. 2). Thus, for instance, the *Atlanta Journal and Constitution* introduced its "Access Atlanta" service in the Bell South's TranstexT Universal Gateway in 1990. For $6.95 per month, subscribers had access to, among other things, news, classified ads, a library of movie reviews, and communication capabilities such as electronic mail and chat rooms ("Videotex debuts in Atlanta," *Editor & Publisher*, October 20, 1990, p. 30).

One indicator of this growth in electronic publishing was the emergence of several research and development efforts in an industry not used to investing money in this type of activity. According to the company's director of new media development, Roger Fidler, in 1992 Knight-Ridder created its Interactive Design Lab with the goal of assessing "the implications of emerging technologies and their impact on newspapers" (Rosenberg 1992, p. 26). A year later, MIT's Media Lab unveiled its research initiative "News in the Future," which had several newspaper companies and other firms as sponsors. This was an unusual development for the industry. Frank Hawkins, a spokesman for Knight-Ridder, said "I don't think anything like this has ever happened in the newspaper business before" (Garneau 1993, p. 13). According to Jim Willis (1994, pp. 109–111), "by mid 1992, the newspaper industry seemed to be entering a new era . . . probing new types of media services, and laying out huge sums of cash on research and development; even cooperating with the competition if it seemed expedient."

This trend accelerated substantially around 1994. "The number of U.S. and Canadian publishers producing or soon to launch online versions of their papers more than doubled since 1993," wrote a reporter reviewing technology developments for *Editor & Publisher* in January 1995 (Rosenberg 1995, p. 59). This acceleration can also be seen in attendance at specialized industry conferences. For instance, 600 people took part in "Interactive Newspapers '94," more than twice as many as in the previous year ("Attendance booms at interactive newspaper conference," *Editor &*

Publisher, February 19, 1994, p. 21), and a few months later more than 500 people went to the annual Newspaper Association of America's "Connections" conference, three times as many as in 1992 ("Connections forum may separate from Nexpo," *Editor & Publisher*, July 2, 1994, p. 31).

This growth acceleration coincided with two broader developments having to do with the Internet. First, the 1992 Clinton-Gore campaign's "information superhighway" rhetoric, and its subsequent materialization in the National Information Infrastructure program after Clinton and Gore took office, captured people's and the media's imagination. Campbell-Kelly and Aspray (1996, p. 299) have argued that "the extraordinary news coverage of the National Information Infrastructure in the early 1990s, on television and in the popular and serious press, caused a boom for the Internet." According to Campbell-Kelly and Aspray, by 1992 "the number of hosts exceeded 1 million for the first time, the following year it exceeded 2 million, and the year after that there were 3.8 million, a number that was growing by 1 million every quarter" (ibid., p. 299). The second development was the transfer of the Internet's administrative oversight and financial support from public to private hands and its opening up fully to commercial uses. This was a process that began in the early 1990s and ended in the spring of 1995 (Abbate 1999).[22] According to Ceruzzi (1998, p. 296), "as recently as 1992, Internet users were about evenly distributed among governmental, educational, military, net-related, commercial, and nonprofit organizations. . . . By 1995, commercial users overwhelmed the rest, and the phrase 'X dot com' [had] entered our vocabulary." Thus, Abbate (1999, p. 199) has suggested that "with privatization, the Internet opened up to a much larger segment of the American public," and that "commercial online services could now offer Internet connections, and the computer industry rushed into the Internet market with an array of new software products and services."

It should then come as no surprise that, caught in the middle of these rhetorical and policy developments, many newspaper people imagined the Internet and related technological changes to be tied to dramatic transformations in their own industry. For instance, in a statement made before shareholders assembled at Richmond Newspapers' suburban Virginia production plant, the chairman and CEO of Media General Incorporated, J. Stewart Bryan III, said that this new information infrastructure was "doing for the communications industry what the industrial revolution did for manufacturing 200 years ago" ("Digital revolution," *Editor & Publisher*, June 17, 1995, p. 19). This centrality of information technologies in the future of the industry is wonderfully captured by a cartoon in which the computer is portrayed not only as the cause of actors' concerns but also as the source of potential solutions (figure 2.4).

The idea that new media would inevitably transform newspapers was coupled with a strong sense of uncertainty regarding the future of the industry. Rather than leading to a wait-and-see attitude, this sense of uncertainty usually seemed to trigger action.[23] This was forcefully manifested by Maxwell King, editor and executive vice-president of the *Philadelphia Inquirer*, in an analysis he wrote after attending a 1995 conference on new media organized by Harvard University's Nieman Foundation. "With virtually everything about this new game uncertain . . . there may be only one immutable rule: If you want to be sure you can play later, you must play now. No communications company can afford to sit out and hope to catch up." ("Three views of the conference," *Nieman Reports* 49, 1995, no. 2: 5, 67–72).

In the midst of this uncertainty, one widespread argument was that print newspapers would have to become "information companies." Thus, at the Newspaper Association of America's 1994 "Nexpo" conference, Joe Hladky, president and publisher of the *Cedar Rapids* (Iowa) *Gazette*, said:

Figure 2.4
A cartoon about computers and the future of newspapers. Source: *Editor & Publisher*, March 19, 1994, p. 4. Cartoon by Douglas Borgstedt.

"The mission of our company states that we want to be the information provider of choice through a dynamic mix of products and services. . . . We don't identify the medium; that choice should be the consumer's." (Case 1994, p. 13) This rhetoric probably contributed to maintaining developments targeted at delivery vehicles other than the personal computer, despite the growing interest in online services. Although audiotex services were up and running through the 1990s, and several of them were mildly profitable (something that could not be said about the online alternatives), perhaps nothing illustrates its demise as a center of attention better than the fate of the annual "Talking Newspaper" conferences. In 1990, the Audiotex Group, a consulting firm, organized the first conference, which focused exclusively on voice information systems and attracted about 40 people. Three years later, the conference had become "Newspapers and Telecommunication Opportunities: Voice, Fax and Online Services." By 1994 it was renamed "Interactive Newspapers," which remained unchanged through the end of the decade. The suggestive subtitle of the 1994 conference was "The Multimedia Mission."

In addition to audiotex, newspapers also tinkered with CD-ROM, fax, portable digital assistants, and bulletin-board services during the first half of the 1990s. For instance, *USA Today* offered CD-ROM "multimedia time capsules," and some fax newspapers succeeded during this period, such as the *New York Times*'s TimesFax, which was being delivered to 150,000 people in 53 countries by 1995 (Thalhimer 1994; "Timesfax now on World Wide Web," *Editor & Publisher*, March 4, 1995, p. 36). Wayne Danielson and his colleagues at the University of Texas worked on expanding the capabilities of Apple's Newton portable digital device for wireless transmission of news (Maher 1994).[24] However, a major blow to this area came in July 1995 when Knight-Ridder closed its Interactive Design Lab, which was heavily involved in portable flat-panel technology. According to CEO Anthony Ridder, the reason for this move was to "concentrate our resources on the Internet and online services" (Webb 1995a, p. 32). The mid 1990s also saw less interest in bulletin-board services, once home to such pioneering efforts as StarText.[25]

Despite these alternative developments, the bulk of activity in the period 1992–1994 took place in relation to online services, most notably America Online, CompuServe, Prodigy, and Ziff Interchange, as opposed to other variants such as bulletin-board services and the then nascent World Wide Web. Thus, that year Prodigy carried online editions of the *Atlanta Journal and Constitution*, the *Los Angeles Times*, *Newsday*, the *Milwaukee Journal*, the *Milwaukee Sentinel*, and the *Providence Journal-*

Bulletin; America Online carried those of the *San Jose Mercury News*, the *Chicago Tribune*, and the *New York Times*; and CompuServe carried that of *Florida Today* ("*N.Y. Times* launches edition on America Online," *Editor & Publisher*, June 25, 1994, p. 117; "Journal/Sentinel to offer Wisconsin on Prodigy network, *Editor & Publisher*, August 27,1994, p. 29; "Providence on line," *Editor & Publisher*, October 15, 1994, p. 35; Rosenberg 1994b). In addition, Ziff Interchange prepared the launch of the *Washington Post*'s electronic service (Heilbrunn 1994).

Related developments included Columbia University's setting aside $5 million to build the Center for New Media in its Pulitzer Graduate School of Journalism, the creation by six newspaper chains[26] of an alliance called PAFET (Partners Affiliated for Exploring Technology), and the launch of the "Initiative for Newspaper Electronic Supplements" by IFRA, the global newspaper and media technology association ("Columbia plans multimedia center," *Editor & Publisher*, August 6, 1994, p. 3; Fitzgerald 1994; Rosenberg 1994a). Another sign of the coming of age of online papers was that during a 1994 strike at the *San Francisco Chronicle* and *San Francisco Examiner* between 10 and 30 stories were available daily on The Gate, the two papers' joint online publication ("S.F. papers were on Internet during the strike," *Editor & Publisher*, December 3, 1994, p. 23).

Things moved even faster and, most important, in a new direction circa 1995 when the web took center stage. There was still significant activity in the online service front. For instance, the *Austin American-Statesman*, the *Dallas Morning News*, the *Houston Chronicle*, and Gannett Suburban Newspapers in New York began publishing electronically on Prodigy, and the *Washington Post* and the *Minneapolis Star Tribune* went live with Ziff Interchange and *USA Today* with CompuServe ("Austin daily goes online with Prodigy," *Editor & Publisher*, March 25, 1995, p. 47; "Gannett's suburban N.Y. papers to go online via Prodigy," *Editor & Publisher*, February 4, 1995, p. 33; Giobbe 1996; "USA Today makes online debut," *Editor & Publisher*, May 20, 1995, p. 35; Webb 1995c). But 1995 was the year that most American newspapers began discovering the web. Some newspapers had published on the web before 1995,[27] but they were only a few of the 175 U.S. papers that, according to the Newspaper Association of America, had web sites by the end of that year ("Number of papers with online edition tripled," *Editor & Publisher*, February 24, 1996, p. 39). As newspapers moved to the web, so did related services. For instance, celebrating the centennial of the comic strip, United Media launched a site, and Tribune Media Services created WebPoint, a suite of syndicated online products (Astor 1995).

The web was created in 1990, but its popularization did not happen until the National Center for Supercomputing Applications at the University of Illinois officially released the first graphical browser, Mosaic, in November 1993. That month 40,000 copies were downloaded. According to Abbate (1999, p. 217), "once Mosaic was available, the system spread at a phenomenal rate. In April of 1993 there had been 62 web servers; by May of 1994 there were 1,248." Scholars have argued that this quantitative growth generated qualitative changes. For instance, Sproull (2000, p. 260) has dated the start of the "third era" of household computing in 1994, defined by the connection of these machines to the Internet and the web. Chandler (2001, p. 174) has taken 1996 to be the "cutoff date" in the establishment of the "infrastructure for the Electronic Century." Concerning electronic publishing, Thorson, Wells, and Rogers (1999) have pointed out that the web became the dominant information environment for new-media advertising in 1995, and Beamish (1997, p. 142) has argued that for British local papers, by that time, "it was becoming clear that the Internet was both the short- and medium-term way forward."

It is useful to contrast the following episodes to gauge the rapid popularization of the web among newspapers. In the spring of 1994, at the Nieman Foundation's first conference on new media, when *New York Times* technology reporter John Markoff mentioned Mosaic, the session's moderator interrupted and asked him to explain what that tool was ("What skills does the journalist require to take advantage of new technology?" *Nieman Reports* 48, 1994, no. 2: 19–25). A year later, things were already changing. An editor at Columbia Journalism Review wrote: "I've been spending a lot of time lately on the World Wide Web. If you haven't heard much about it yet, you will soon. Hundreds of newspapers . . . are racing to establish presences on this rapidly growing . . . section of the Internet." (Hearst 1995, p. 63) Just a few months later, "web" had already become a common word in newspaper organizations across the United States. "By the end of 1995," according to Fitzgerald (1996, p. 14), "even small weekly newspapers had home pages." In February 1996, *Editor & Publisher* devoted its annual new-media supplement to the web. The introductory article stated: "The trickle grew into a stream, the stream has swollen into a river, and nobody knows when the flood of newspapers into computerized information services will peak. . . . Overwhelmingly, the highway these information services are traveling is the World Wide Web. . . . Already, newspapers are beginning to exit the proprietary toll roads— such services as America Online or Prodigy." (Garneau 1996, p. 2i)

Some online services reacted to the popularization of the web by trying to mediate between their consumers and the new environment. For instance, in an ad in *Editor & Publisher*, Prodigy claimed to be the first major online service to provide access to the web, which it hoped could benefit its efforts in the online newspaper field (figure 2.5). Barry Kluger, senior vice-president of communications for Prodigy, commented: "Prodigy is not getting out of the newspaper business. . . . We are simply becoming webcentric and meeting the needs of newspapers. . . . Rather than fighting the migration to the Internet, we are going to be holistic with it." (Cohen 1996, p. 40) Unfortunately for Prodigy, this strategy did not succeed. A signal of its failure was the sale a year later of Prodigy's online classified advertising system to Thomson Newspapers, which then used it in many of its properties' web sites as well as in those of other firms' newspapers (Cohen 1997). *Editor & Publisher*'s new-media columnist reflected about the evolution of events as follows: "If 1995 made one thing clear, it was that the World Wide Web is the online publishing platform of choice." He added that "nearly 90 percent of all online newspaper services worldwide are accessible via the web. Only in the United States are some newspapers operating services on proprietary commercial services, but their numbers have remained flat." (Outing 1996, p. 5I)

Settling, Settlements, and Settlers

After more than a decade of exploring various electronic alternatives—many of which never ceased to be pursued, albeit in a comparatively minor fashion—American dailies progressively settled on a particular information environment. In the mid 1990s the delivery of content and applications to personal computers connected to the web achieved a dominant status.[28] As Molina (1997a, p. 208) has put it: "De facto, the web has become the prime arena for multimedia newspaper developments in the near and medium-term future." How are we to account for this narrowing down of the electronic publishing alternatives explored? Research in sociology, in history, and in the management of technology has shown that the passage from many options to one dominant choice is not the logical outcome of some technical superiority, but results from an array of social processes that coalesce action around the "winner" and discourage investment in the "losers."[29] Within this broad perspective, I make sense of the present case's dynamics by resorting to the notion of "settling," exploiting the word's connotations of settling a dispute, settling in, and the actors as settlers.

Figure 2.5
Prodigy gives access to the World Wide Web. Source: *Editor & Publisher*, February 4, 1995. © Prodigy Communications Corporation. All rights reserved. Advertisement courtesy of SBC Intellectual Property and Prodigy Communications Corporation.

First, settling a dispute builds on the notion of "closure" central to the social construction of technology model.[30] As defined in the model's original formulation: "Closure in technology involves the stabilization of an artifact and the 'disappearance' of problems. To solve a technological 'controversy' the problems need not be solved in the common sense of the word. The key point is whether the relevant social groups see the problem as being solved." (Pinch and Bijker 1984, pp. 426–427) Initially, closure was conceived as a rather static phenomenon: once achieved, it was seen as very difficult to undo.[31] In a review of research in the social construction of technology tradition, Bijker (2001, p. 15,524) has stated that closure "highlights the irreversible end point of a discordant process in which several artifacts existed next to one another." However, other scholars have recently espoused a more dynamic view. For instance, Kline (2000) has shown what happens when consumers reinsert interpretive flexibility into a "closed" artifact, and Pinch (2001, p. 398) has argued that it is "more useful to see closure as something that is continually in operation." My use of "settling a dispute" follows this more recent work by combining connotations of stability and change.[32] Similar to disputes in other areas of life, disputes over technical matters are episodes in ongoing relationships. Thus, settling a particular dispute endows the chosen option with a certain degree of hardness that facilitates further developments. However, by being part of a dynamic sequence of events, future contingencies may lead actors to reconsider existing settlements, thus challenging their obduracy—an aspect that the notion's second and third meanings, developed below, reinforce.

How did actors settle the dispute of online services versus the web circa 1995? Seeking social projections of technical features does not provide a satisfactory explanation. Had they chosen the web because of its superior multimedia capabilities, it would be hard to explain why they took limited advantage of them later on. Similarly, had they preferred the web because of its distinctive hyperlinking qualities, there would have to have been a broad change in the linear and self-contained authoring culture of the industry, something of which there were no major signs. What about the possibility that on the web newspapers had more freedom to build their own artifact than on someone else's online service? That factor may have influenced the election of one online service over another. For instance, a *Washington Post* spokesperson said her company chose to publish with Ziff-Interchange because it was "the only on-line service that [among other things] enables the publisher to create its own 'look and feel' for the electronic edition [and] preserves the company's direct business relationship with Post readers" (Rosenberg 1994c, p. 40). Delphi was one online service that tried to capitalize on this preference (figure 2.6). By extension, the web could have been more attractive to actors used to an autonomous product culture. But this alone could not account for their rapid migration from online services after 15 years of undertaking projects in environments they controlled only to a very limited extent. So, at the most, this could only be a supplementary factor.

Rather than multimedia-, hyperlinking-, and autonomy-related issues, it seems that what most contributed to settling the dispute was the perception among key decision makers that the web was becoming the preferred environment for users. Two high-profile moves from online services to the web—the *San Jose Mercury News* from America Online and the *Los Angeles Times* from Prodigy—illustrate this matter. When Mercury Center launched its web edition in February 1995, Bill Mitchell, its director of electronic publishing, said: "We . . . want to be on the Internet. A year ago, we decided that it was clear that more and more people are making the transition to a commercial Internet." (Conniff 1995, p. 3) Eight months later, when the *Los Angeles Times* announced the launch of its web site for early 1996, it also stated that that would conclude TimesLink, its service on Prodigy ("TimesLink to move off Prodigy, onto the web," *Editor & Publisher*, October 28, 1995, p. 45). Echoing Bill Mitchell's words, Richard Schlosberg III, publisher and CEO of the *Los Angeles Times*, said: "In the rapidly evolving online world, it is important to go where the customers—both users and advertisers—are going." ("*L.A. Times* exits Prodigy, for now," *Editor & Publisher*, December 2, 1995, p. 37) Paraphrasing Prodigy's Barry Kluger, one might say that if users were becoming "webcentric" then choosing online services could entail leaving the playing field to an already crowded and growing competitive set. Furthermore, in a survey of online newspaper editors conducted in early 1997, Peng, Them, and Xiaomin (1999) found that the top reason for choosing the web over online services was the availability of large audiences worldwide. This 57 percent was more than double the next stated reason, which was ease of publishing. Thus, in the social construction of technology model's parlance, the mechanism[33] that settled the dispute could be termed "closure by perceived user behavior." The perception by decision makers that the process was unfolding in a specific direction is analytically more central than what users were in fact doing (something that was difficult to ascertain in the midst of a rapidly changing situation).

Figure 2.6
Delphi Internet lets newspapers design their own "look and feel." Source: *Editor & Publisher*, April 23, 1994. © Prospero Technologies.

Closure by perceived user behavior is congenial with the larger context in which online newspapers have unfolded, since, as Abbate (1999, p. 4) has argued, "the culture of the Internet challenges the whole distinction between producers and users." In such a context, the boundary between technological development and economic consumption gets blurred, as artifacts evolve almost seamlessly from the laboratory to the marketplace. Rather than reopening an already hardened online paper, which is the more usual account of users as agents of technical change,[34] perceptions of user behavior contributed decisively to its initial closure as a web artifact. This type of closure mechanism also helps account for the role of market forces in the development of artifacts, a relatively underdeveloped area in the social construction of technology model.[35]

Settling the dispute was recursively linked to building web sites: the more sites were built, the more the dispute seemed settled, and the more settled the dispute seemed, the more sites were built.[36] This leads us to the notion of "settling in" as a development-oriented activity crucial to both the stability of existing settlements and their prospective evolution. This was what newspaper people did: they became settlers, establishing a presence in an unknown territory, mapping its contours, bringing along their old products, creating new ones, and offering their wares to old and new customers. Much like settlers in "real" territories, they were constantly faced with new challenges for which their old behavioral and symbolic repertoires were of only limited utility. Gordon Borrell, an executive with the online publishing firm InfiNet, put it this way (1995, pp. 21TC, 26TC): "The online world is truly the Wild West. Only this time there are a lot more Indians and they aren't about to give up the land. . . . While we shovel all that 'important' news onto our World Wide Web sites, the most popular place to visit on our web pages is 'The Cool Site of the Day,' which merely points to an interesting location that one of our programmers found out in Internet land."

The "conquest" imagery in Borrell's statement points to the political dimension involved in establishing settlements and becoming settlers. Online newspapers did not settle in an empty space. On the contrary, before the web's popularization circa 1995, the Internet had already evolved a set of communication discourses and practices that were sometimes in conflict with traditional media's modus operandi. This was especially the case, clearly expressed in Borrell's quote, of the tension between the preference of media organizations for unidirectional communication and the multi-directional patterns enacted in information infrastructures such as bulletin-board services and Usenet. This tension was explored above in the Viewtron case and will be examined again in subsequent chapters.

The emphasis on human agency in the forward-looking aspect of settling contrasts with two related notions that have emphasized the technical shaping of events once the passage from many to one alternatives occurs: "dominant design" (Abernathy and Utterback 1978) and "momentum" (Hughes 1969). For instance, Tushman and Murmann (1998, p. 244) have argued that "where social, political and institutional forces shape technological progress prior to the dominant design, technology drives subsequent technical evolution after the dominant design." Regarding momentum, Hughes has suggested that "mature systems have a quality that is analogous . . . to inertia in motion" (1987, p. 76), and that "as they grow larger and more complex, systems tend to be more shaping of society and less shaped by it" (1994, p. 112). In contrast, the indeterminacy involved in establishing settlements and becoming settlers emphasizes the centrality of human agency in the process of innovation after the actors settle in a specific alternative. Thus, building upon but also expanding existing scholarship, the notion of settling helps to make sense of the processes whereby the web became the preferred information environment for online newspapers circa 1995. How things unfolded in the years immediately following these settlements will be discussed in chapter 3.

Newspapers' Culture of Innovation

From the early 1980s to the mid 1990s, American dailies tinkered with an array of alternatives to print, from stand-alone videotex systems to web editions. I have argued that the 1980s was a decade of exploration of multiple technical, editorial, and commercial options. While newspapers continued to explore these options during the first half of the 1990s, they progressively narrowed down their efforts around products delivered to personal computers connected to online services until they finally settled on the web circa 1995.

Beneath this shift from exploring to settling lies a culture of innovation marked by a combination of reactive, defensive, and pragmatic traits. First, by 'reactive' I mean that quite often newspapers acted only after it seemed evident to key decision makers that relevant technical and social developments had a reasonable chance of taking hold, rather than more proactively trying to take advantage of them earlier in the game. For instance, the videotex and teletext endeavors of American dailies followed initiatives that occurred first in other countries. The move to the web happened after the public appeared to show enough signs of strong interest in it and some early entrants, such as Netscape and Yahoo!, had reached significant market success. Second, the word 'defensive' underscores that newspapers were usually more interested in finding out what the new technologies meant for the print enterprise than in more offensively developing new technical, communication, and organizational capabilities. According to Ettema (1989, p. 108), "interest in the technology on the part of newspaper firms . . . was probably more defensive than offensive. . . . The failures of videotex ventures were, then, at worst a mixed blessing for the newspaper industry." For instance, Viewtron officials publicly declared that the project had been initiated to protect Knight-Ridder's print franchises, the industry as a whole fought quite intensely to keep AT&T and the regional Bell operating companies from engaging in electronic publishing, and the main reason for undertaking audiotex efforts was to maintain a paper's position in its community.[37] Third, I use the word 'pragmatic' to underscore that with their nonprint innovations, American dailies were often more interested in the short-term health of the core print businesses than, more idealistically, in projects that seemed more promising with comparatively higher payoffs that could only pan out in a longer term. "Many newspapers," according to Carey and Pavlik (1993, p. 165), "developed videotex services not as a positive step forward but out of fear that videotex might replace their core business. When it became clear that the perceived threat was illusory, they retreated quickly." Thus, the Associated Press–CompuServe experiment was terminated on the basis of a conclusion that videotex did not entail "clear and present danger to newspapers," Viewtron was folded because it was not "likely to be a threat to the newspaper industry in the foreseeable future," and Knight-Ridder closed its research and development facility charged with experimenting with portable flat-panel devices when online services and the web seemed to turn irrelevant all other possible technical scenarios.

In the next chapter, after examining the first 5 years of online papers on the web, I will elaborate further on the implications that this culture of innovation has for understanding how an established media industry appropriates technological developments that both open new horizons and threaten the status quo. But beyond the dynamics of this culture of innovation, this analysis also underscores the heuristic value of a historical perspective in the study of emerging media. The events analyzed in this chapter tell a story of breaks and changes of pace within a background of unfolding transformations. By locating the source and timing of both continuities and discontinuities such an account helps to avoid some misunderstandings about the efforts by American dailies in the area of consumer-oriented alternatives to print. Most notably among these misunderstandings is a belief, quite pervasive in both academic and popular discourse, that the creation and growth of online newspapers on the web was some sort of revolutionary occurrence and without any roots in the past. We find expressions of this belief even in the work of otherwise historically inclined scholars. For instance, in their illuminating study of the development of the news form from the American Revolution to the present, Barnhurst and Nerone (2001, p. 284) have commented about newspapers' appropriation of the web as follows: "Throughout history, change at newspapers has come deliberately and often with great reluctance. . . . In the history of news form, the leap onto the web came precipitously." In contrast, my account has painted a different picture: although settling on the web meant a qualitative break from, and happened faster than, prior alternatives to print, it was a far more evolutionary process influenced by a history of tinkering with multiple forms and many facets of consumer-oriented electronic publishing.

3

Hedging: A Web of Challenges in the Second Half of the 1990s

The second half of the 1990s was a period of effervescence about anything related to the web, and American dailies were no exception: they poured human, financial, and symbolic resources into their nonprint endeavors with an intensity the industry had not seen before. In these endeavors, online newspapers simultaneously undertook three kinds of information practices. First, newspapers repurposed content by utilizing in an almost identical fashion on their web sites material originally developed for their print editions. Second, they recombined information by taking print content and increasing their usefulness on their web sites through the addition of technical functionality or related content from other sites or both. Recombination of practices included personalized or customized editions, new sites pulling together vast amounts of news and database information on a particular topic such as city guides, sites linking similar content from many online newspapers such as nationwide classified ads, and archives of past editions. Third, dailies created original content by taking advantage of the unique capabilities of the web. This type of practice included constant updates on breaking stories during the day, special multimedia packages of major events, new sections developed exclusively for their web sites, and user-authored content.

After exploring an array of nonprint options in the 1980s and settling on the web in the mid 1990s, what resulted from this multiplicity of information practices in the second half of the 1990s can be characterized as a form of hedging. Hedging emerged as a response to uncertainty in a volatile operating environment: newspapers spread risks by moving in many and often counterbalancing directions.

In chapter 2, I argued that American dailies, in their pre-web endeavors, had exhibited a culture of innovation marked by reactive, defensive, and pragmatic traits. Such a culture of innovation was also prevalent in their web efforts, which helps to account for the seemingly contradictory mix of practices enacted in this period of hedging. More generally, after two decades of tinkering with nonprint delivery vehicles influenced by this culture of innovation, the overall consequence was twofold. On the one hand, newspapers often appropriated new technologies with a somewhat conservative mindset, thus acting more slowly and less creatively than competitors less tied to traditional media. On the other hand, the cumulative transformations should not be underestimated. By the end of the 1990s, online newspapers exhibited a technical infrastructure, nascent communication and organizational patterns, and a suite of products that looked very different from those of a typical print counterpart. It appears that in a relentless pursuit of permanence, newspapers ended up undertaking substantial change.

Hedging

The second half of the 1990s saw explosive growth in online communications. The number of Internet hosts increased from 5.8 million in January 1995 to 72 million in January 2000 (Chandler 2001). In 1997, 37 percent of households in the United States had personal computers, and America Online reached 10 million subscribers (Sproull 2000). By mid 1999, 106 million people in the United States, or 40 percent of the country's adult population, were online (Compaine 2000b). Commercial activity was also on the rise; of particular importance for the newspaper industry, advertising on the web rose from $267 million in 1996 to almost $3 billion in 1999 (Compaine 2000b).

Online newspapers also developed considerably during this period. According to Levins (1997d), the number of online newspapers "more than doubled" during 1996. A Newspaper Association of America tally showed that by April 1998 "more than 750 North American daily newspapers [had] launched online services" (Newspaper Association of America 1998). By July 1999 only two of the 100 largest dailies did not have an online presence (Dotinga 1999). The staffs of online papers also grew. For instance, by July 1997 the Chicago Tribune Interactive Edition employed 80 people, USA Today Online employed 84, the Wall Street Journal Interactive Edition employed 90, and the *Washington Post's* Digital Ink employed 100 (Kirsner 1997a). Related activities such as syndication also grew. According to *Editor & Publisher's* syndication specialist, "in 1995, the wired world of syndication became acquainted with the World Wide Web," and in 1996 "they became very close friends" (Astor 1997b, p. 64). This trend continued in 1997, when sales of syndicated content for web sites increased about 200 percent at United Media and when 20 percent the Universal Press Syndicate's sales were to nonprint outlets (Astor 1997a).[1]

Site traffic also grew substantially, more than doubling every year from 1995 to 1998. A study conducted by the Internet usage auditing firms I/PRO and Media Metrix concluded that "traffic to established media · web sites increased by 130 percent in 1997" ("'Astonishing' growth," *Editor & Publisher*, May 16, 1998, p. 34). As time went by, the traffic in online newspapers began to rival their print circulation. For example, by early 1999 the *San Jose Mercury News* had a daily circulation of 290,000 while its web site, Mercury Center, was getting 100,000 visitors per day, and the editor of the New York Times on the Web estimated that on a typical weekday about 250,000 visited the site of the daily, which had a circulation of more than 1 million (Outing 1999b).[2] More generally, a survey of 3,184 adults conducted by the Pew Research Center for the People and the Press found that the proportion of people getting the news online at least once a week rose from 4 percent in 1995 to between 15 and 26 percent in 1998 (Noack 1999b). The variation resulted partly from the absence or presence of high-impact stories, which generated peaks of usage. For example, washingtonpost.com served 4.5 million pages the day the Starr Report was released (Stone 1998).

Online newspapers became mainstream as a result of this growth in supply and demand. Thus, the Society for Professional Journalists created the first category for online journalism in its prestigious Sigma Delta Chi awards in 1997 and added two more in 1998 ("Online awards expanded," *Editor & Publisher*, April 17, 1999, p. 20). Institutionalization continued when the Online News Association "opened for business" in the spring of 1999 (Noack 1999a). This mainstreaming was also evident in the experience of software vendors. For example, in 1997 Steve Burns, vice-president of new technology at Gannett Media Technologies, said: "A year ago . . . when you went into a paper, if it even had an 'online manager', that person was usually someone who had been given those duties in addition to their real job and did them halfway. Now, we regularly go into newspaper meetings to find the classified ad manager and ad director at one end of the table in suits and ties and these people in jeans at the other end—the online group. They used to be sheepish and quiet in corporate meetings because they were just overheard. But now we're seeing these online managers rising to positions of real power within these corporations. And, the same online managers are starting to get substantial funding and purchasing authority." (Levins 1997a, p. 45)

The mainstreaming of online newspapers could also be seen in an ad by Hearst Newspapers in which online is portrayed as a publishing environment as important as print. It is also worth noticing that the corporation's logo is engraved in a microchip-like object, signaling the centrality that modern computing has acquired in the whole newspaper enterprise (figure 3.1). And in a move with strong practical and symbolic implications, in 1998 Knight Ridder[3] announced the relocation of its headquarters from Miami, Florida, to San Jose, California. Anthony Ridder, the corporation's chairman and CEO, said "Knight Ridder people simply must be immersed in the kind of futuristic and entrepreneurial thinking

Figure 3.1
Mainstreaming of online newspapers. Source: *Editor & Publisher*, July 19, 1997. Reprinted with permission.

found in Silicon Valley." ("New KR HQ: Silicon Valley," *Editor & Publisher*, May 2, 1998, p. 16) Randy Bennett, vice-president for electronic publishing at the Newspaper Association of America, described the mood at the 1998 "Connections" conference as follows: "The days of evangelism, hand-wringing and talk of experimentation are really over. . . . Online newspaper publishing is now an established business and people aren't looking to be convinced anymore. They came here this year for hard, detailed information about the logistics of online publishing." (Levins 1998b, p. 20)

What did online newspapers consist of in this period of growth and mainstreaming? As I briefly described in the introduction to this chapter, as settlers ventured into the new territory, they hedged. Actors continued exploring and settling, but their initiatives became less experimental and more competitive.[4] To account for these dynamics, in the remainder of this section I will first divide key developments into three types of information practices: repurposing, recombining, and recreating. Then, I will analyze how this multiplicity of information practices amounted to a form of hedging that newspapers enacted in response to an uncertain and changing environment.

Repurposing

'Repurposing' and 'shovelware' were terms often used to refer to the common practice of taking information generated originally for a paper's print edition and deploying it virtually unchanged onto its web site. In the June 1996 issue of *American Journalism Review* we read: "Most of the newspapers on the Net are producing 'shovelware,' print stories reproduced on web pages, with few changes other than key words painted hypertext blue that offer readers links to stories with greater depth." (Pogash 1996) Research has shown that repurposing was the dominant information practice not only of American but also of Asian and European online papers during this period.[5]

Despite its apparent simplicity, repurposing had some complex implications. Such was the case of the relationships between newspaper organizations and freelancers. Historically, newspapers had tended to give freelancers secondary publishing rights, letting them publish contributed stories in other outlets. However, when the web began to become mainstream, several publishers became reluctant to continue to permit them to do so and asked freelancers to sign agreements transferring to them copyrights of their products published electronically or in any other form. For instance, on July 20, 1995 managers at the *New York*

Times circulated a memo to department heads and section editors stating that freelancers who did not sign such an agreement would "no longer be published in the newspaper" (Grunfeld 1996, p. 10). A cartoon in *Editor & Publisher* illustrated the freelancers' situation (figure 3.2). Seeking to counter this trend, twelve freelancers, with the support of the National Writers Union, sued the *Times* and some other publishers for infringing their copyrights (Savell 1996). Dan Carlinsky of the American Society of Journalists and Authors stated the writers' position: "When I create a piece as an individual contractor and not as an employee, it's mine and I may license it to others, not the newspaper." (Anderson 1997, p. 53) Kenneth Richieri, assistant general counsel for the New York Times Corporation, disagreed: "When no agreement exists, can a publisher include freelance pieces in an electronic version? . . . We say that even if we've never talked about microfilm, we can put articles on microfilm. The same rule applies to CD-ROM, Lexis-Nexis and other media." (ibid.) In the summer of 1997, U.S. District Court Judge Sonia Sotomayor ruled that the publishers had not violated freelancers' rights under the federal copyright law (Noack 1997); however, the plaintiffs later won an appeal. As the 1990s ended, this litigation was ongoing. On June 25, 2001, the U.S. Supreme Court ruled in favor of the freelancers (Greenhouse 2001). Regardless of its final outcome, the lawsuit is significant for present purposes because it shows how seemingly innocuous

Figure 3.2
Freelancers and the electronic republication of their work. Source: *Editor & Publisher*, January 24, 1998, p. 7. © Steve Greenberg. Reprinted with permission.

repurposing affected newspapers' established practices as they moved to the web.

As time went by, the dominance of repurposing began to recede. This happened partly because automation of the whole process freed resources for other types of activities and partly because online newspapers undertook other information practices to contend with companies developing new web-based products, which were either competing with the newspapers' franchise or could eventually become a new and attractive source of revenue.

Recombining

In addition to repurposing, newspapers also recombined existing content on their online editions and related sites.[6] Here 'recombining' refers to information practices that took some content originally generated for the print edition and substantially increased its utility on the web by supplementing it with new content or with similar content from papers of other geographic locations, and/or adding new functionality to the ways in which that content could be accessed, manipulated, and, in general, used.

One type of recombination had to do with the customization of an otherwise generalized product, realizing the vision of "The Daily Me" articulated by researchers at MIT's Media Lab since the mid 1980s and embodied in their 1993 "FishWrap" prototype.[7] Rather than present the same editorial content to all users, some papers began providing individual users stories about only those things they had previously declared were of interest to them. Customization was not confined to editorial content, but was also used in the delivery of ads.[8] One online paper that took advantage of this commercial side of customization was the New York Times on the Web, charging advertisers a premium if they requested their ads to be delivered to specific groups of users. Although most of the site was accessed free of charge, the Times on the Web required users to complete a registration form before being given access. These data, combined with a record of each registered user's behavior on the site, were processed in real time with the aid of software developed in house. "This constant tracking allows us to modify campaigns in flight," stated Peter Lenz, research director for the New York Times Electronic Media Company (Sullivan 1999, p. 44). For example, in a campaign for a telecommunications company, the online paper "delivered ads only to those living in neighborhoods near the teleco's retail stores in one metropolitan area" (ibid.).

A second form of recombination emphasized the provision of large amounts of content and an array of applications on specific topics, enabled by exploiting the virtually unlimited "news hole" of the online environment and its interactive capabilities. As opposed to the somewhat static and spatially constrained print environment, which gives its readers a "horizontal"[9] overview of the every day's main occurrences in most important areas of society, these "verticals" provided a seemingly unending stream of information and services focused on a particular matter. One type of vertical that attracted significant attention and activity was the "online city guide," dubbed "the hottest new real estate in cyberspace" in a 1996 *Columbia Journalism Review* article (Houston 1996). It provided an array of options, including restaurant reviews, searchable event calendars, electronic yellow pages, and interactive maps. The "online city guide" arena became very competitive during the second half of the 1990s, when America Online, Microsoft, and Yahoo launched products in the largest metropolitan markets. This process included acquiring content usually from established information providers and hiring journalists to develop new articles (Flynn 1998). This market became very crowded in some cities; Boston, for instance, had eight different guides by the end of 1997 (Kirsner 1997c).

Another form of recombination had to do with putting together a specific type of content from various newspapers throughout the country and adding services ranging from search capabilities to electronic mail reminders. Like verticals, the focus was on a particular type of information. However, in contrast with verticals, the goal was to add value by providing a network of similar information in many cities and towns. Thus, these networks aimed to extend the reach of online newspapers' products beyond the traditional local boundaries of their print counterparts. Sometimes these networks tied together various properties of the same newspaper chain; on other occasions, they were the results of alliances and partnerships among firms. In the second case, these alliances and partnerships were often among newspaper companies more accustomed to competing than to cooperating. For instance, when five newspaper chains[10] bought the online service AdOne Classified Network in the spring of 1999, George Irish, president of Hearst, touted such partnerships as "becoming common" and foresaw the day that "every newspaper in the company will participate together in a national site so the strength of the entire market is evident in one location online" (Liebeskind 1999).

Unsurprisingly in view of the economics of the industry, strong embodiments of these networks of sites emerged in the area of classified ads, a critical franchise that was seen as threatened by new online competitors.[11] Two of the most significant endeavors in this area have been CareerPath.com and Classified Ventures. CareerPath.com, an employment site, was launched in October 1995, when the *Boston Globe*, the *Chicago Tribune*, the *Los Angeles Times*, the *New York Times*, the *San Jose Mercury News*, and the *Washington Post* pulled their help-wanted ads together on a single web site (Webb 1995b). The site grew rapidly; by mid 1996 there were 26 papers posting nearly 150,000 jobs at any time, which were being searched 4.5 million times per month (Liebeskind 1997). Both the number of contributing papers and the additional services offered had expanded vastly less than 2 years later. Classified Ventures was initially a partnership of Times Mirror, the Tribune Corporation, and the Washington Post Corporation concentrating on automobile and real-estate classifieds. This enterprise also grew quickly, and by August 1998 it had more than 150 affiliate papers, reaching 34 of the top 50 markets in the United States (Levins 1998c). This growth was accompanied by a discourse that countered the print-newspaper culture of local autonomy. For instance, when Knight Ridder joined Classified Ventures in early 1998, Bob Ingle, who was in charge of Knight Ridder's new-media efforts, contrasted local print fortresses with global online armies: "Fortresses are about geography, defending your home base. That's the old world. The Internet transcends geography. . . . The days of building local fortresses are over. We're building an online army." (Stone 1998, p. 32)

Another form of recombination had to do with turning the "morgue" (containing old newspaper articles and used mostly in house) into an archive that could be searched on the web. How far back these collections went and how much users were charged varied from site to site. What did not vary was that "breathing new life" into a morgue became a logical extension of print newspapers as they moved into the online environment (figure 3.3). Publicly available digital libraries proved financially rewarding for at least some of the online newspapers that developed them.[12] For instance, in January 1998 *USA Today* began charging for use of its online archives because they were the "single most requested product" on its site ("*USA Today* archives," *Editor & Publisher*, January 17, 1998, p. 28). Knight Ridder moved even more aggressively in this direction, developing NewsLibrary, a new product aimed at generating revenue from building a network of archives. NewsLibrary offered users access to the old stories of dozens of newspapers and had, for instance, 25,000 article downloads during the 1997 Christmas season (Neuwirth 1998a).

Figure 3.3
Newsday.com creates a publicly available digital library. Source: *Editor & Publisher*, October 26, 1996. © 1996 Newsday, Inc. Reprinted with permission.

Recreating

In addition to repurposing and recombining, online newspapers engaged in an array of practices I call 'recreating', which basically consisted of providing their users with content developed primarily, if not exclusively, for their sites. Although the focus was on generating something new, I use the prefix 're' to emphasize the fact that these practices drew partly from symbolic, behavioral, and material repertoires already existing in media and computing circles—for example, writing genres, video editing procedures, and animation techniques. Although recreating constituted a minority of the information practices undertaken by online newspapers, it is worth noting that these practices experienced a steady increase.[13] No data on this increase were available, but *Editor & Publisher*'s new-media columnist put it this way: "Online-exclusive content being produced by online news sites is growing handily." (Outing 1999a)

Among the manifestations of recreation practices were the updates that some newspapers regularly included on their sites during the day.[14] That is, in addition to the more repurposed-like form of updates, e.g., by featuring a "live" raw wire feed, a web-enhanced wire, or wire copy edited by online staffers, some online papers provided reports during the day originally produced by either their own personnel, reporters at the print newsroom, or both. Because of their more direct relevance to users, most such updates tended to concentrate on either business or local news, and, in some cases, on local business news. For instance, every working day at 4:30 P.M. the *Chicago Tribune*'s print business desk produced a "first edition" of the next day's business news for the paper's site. Owen Youngman, the *Tribune*'s director of interactive media, was quoted as follows: "The late afternoon is a peak time for Internet use, [so] we now provide serious information searchers compelling content in advance of its print publication." ("Tomorrow's news now," *Editor & Publisher*, September 13, 1997, p. 34)

Another embodiment of recreation practices were the "specials." They usually consisted of an in-depth look at a phenomenon or matter of particular attractiveness, from a major sports event to a salient health-care issue. They also permitted newspapers to experiment more intensely with media, and their technologies, that they did not use very often, like audio, video, and computer animation. One highly acclaimed special was the web version of the *Philadelphia Inquirer*'s "Blackhawk Down," a series of articles on the 1993 battle between U.S. soldiers and Somalian rebels in Mogadishu (figure 3.4).[15] The series was published in the paper in 30 installments starting on November 16, 1997. In addition to that version, Knight Ridder, the *Inquirer*'s parent company, produced a one-hour documentary aired on the Public Broadcasting System, a book, and a 30-part hypermedia package featured on Phillynews.com, the joint web site of the *Inquirer* and the *Philadelphia Daily News*.[16] The making of this project took more than a year and included the participation of personnel from several Knight Ridder units. The online version featured text, photographs, video, audio, original graphics, animated maps, twenty question-and-

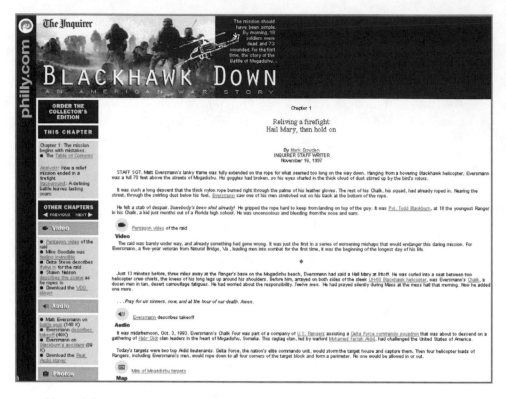

Figure 3.4
The homepage of "Blackhawk Down." Source. Philly.com. © Knight Ridder/
Tribune Media Services.

answer sessions with the main writer, a forum, and its own search engine.
According to Gary Farrugia, Knight Ridder Video's editorial director, the
site was "a new kind of creature [that demonstrated] the amazing potential for online journalism to combine the dramatic narrative capability of
TV, film, and radio with the depth and breadth of factual content afforded
by print" (Levins 1997c, p. 23). This version proved very popular with
users and industry peers. Shortly after its launch, it was getting 28,000
page views per day. A few months later, it received the 1997 EPpy Award
for Best Special Section in a Newspaper Online Service ("Web interest
high," *Editor & Publisher*, February 7, 1998, p. 6; Levins 1998a).

   In view of their technical sophistication, the production of specials usually confronted online newsrooms with challenges not heard of in their
print counterparts: the challenges presented by users' technical heterogeneity.[17] Users connected to the web at different speeds, from diverse

platforms, and by employing various browsers, as well as with different degrees of technical expertise and even wider interests in becoming more proficient. All these factors affected their online experience. This was well known among online newspaper personnel, who confronted the issue when they undertook an initiative involving sophisticated technical, design, and/or media elements. To Steve Yelvington, editor of Minneapolis StarTribune.com, this was "tantalizing and frustrating": "We have some applications developed in Shockwave on our server and we don't use them anymore. They don't work with Internet Explorer 4. There are a lot of exciting and cool things out there, and we'd love to have them. But if they don't work with a reasonably sized subset of our audience, we're not going to use them. The big picture is that this technology is rapidly evolving and content that you author in 1996 may or may not work in a 1998 browser. Most troubling." (Featherly 1998, p. 22)

A third form of recreation consisted of producing original content on a regular basis by personnel at either the print, online, or both newsrooms. Unlike the specials, with their one-time nature, this type of recreation consisted of a steady stream of content. Because of the need to generate a constant flow of new information, and the limited resources usually available, this type of product tended to be less sophisticated than the specials in terms of its use of audio, video, and/or interactive computing capabilities. One particularly aggressive site in this area was washingtonpost.com, especially in its coverage of political news, supplementing the paper's content with original columns, briefs, newsletters, and so on. This in turn gave the site a level of institutional access enjoyed by few online news operations. According to Mark Stencel, the site's politics editor, "to get our Capitol Hill press credentials . . . we had to prove to the standing committee [granting them] that we were doing significant original reporting." (Outing 1999a)

Another manifestation of recreation practices was user-authored content.[18] Although users contribute content to print newspapers in letters to the editor, op-eds, and so on, this has remained a relatively small source of information and one that has been filtered by editors at the print newsroom. On the other hand, online newspapers have seen a growth in quantity and diversity of user-authored content from forums and chat rooms to ranking and reviews to self-publishing.[19] This marked a significant departure from the videotex experience of the early 1980s. Perhaps the most innovative expression of this type of content during the period covered in this chapter was "community publishing." This consisted of an online newspaper giving space and tools to its users so that they could build their own publications within the paper's site. In comparison with non-news sites such as Geocities and Tripod, online newspapers' community publishing efforts had "been slow to take hold—mostly because many publishers still resist the notion that they should be in the business of letting people other than themselves 'publish' news" (Outing 1998b). One of the first initiatives of this nature was by New Jersey's *Bergen Record*, which in partnership with software vendor Koz launched a program that allowed community organizations to create their own online publication within the paper's site. According to Glenn Ritt, the Record's vice-president of news and information, this type of project meant "recognizing that you are a communications company providing an [information] infrastructure"—an infrastructure in which the content is owned by their creators. "We don't want to be the 'publisher,'" said Ritt, "we want to be the 'host.'" (Outing 1998a)

Dealing with Uncertainty

If one examines the result from the weaving of repurposing, recombining, and recreating practices, one sees something remarkably different from the print artifact that has populated the American media landscape for almost three centuries. First, a largely generalized product has turned into one that can easily be customized to every consumer's preferences. Although it is still early to assess the consequences of customization in news and advertisement content, it is undeniable that online papers have deepened a process of unbundling a unitary media artifact that began with the growth of print papers' sections a few decades ago. Second, an entity in which content and form have been partly predicated upon the spatial limitations of newsprint has turned into one of "verticals" with unlimited newshole. The wealth of editorial, advertising, and database information available in, for instance, a well-developed real-estate section of an online paper, ranging from hundreds of news stories about the market as well as particular neighborhoods, to multimedia ads, to detailed records of anything from past transactions to school test scores, is a clear indication of the difference with what is feasible in a print paper. Third, an artifact that distribution costs have usually confined to a place has become simultaneously both micro-local and global. It can feature content such as high school basketball scores that interests only a handful of users while, at the same time, being accessible to migrants located around the globe who want to learn the news of their hometown.

Fourth, an object that often lasts only 24 hours has been made to extend its duration by having past editions as readily available as the latest one, *de facto* turning into a permanently available digital library. Fifth, an artifact produced in mostly fixed cycles has been made more complex by featuring constant updates. This has challenged the division that has marked the news business for the past several decades of television, which informs when breaking news occur, and print, which explains the next day what they mean. Sixth, a product in which information could be communicated only through text and still images has now become multimedia, triggering the possibility of major transformations in the storytelling conventions of news reporting. Seventh, an information architecture dominated by one-to-many linear flows has exploded to include various forms of user-authored content. This has the potential of turning consumers into co-producers and altering the role of journalists as the single source of information available in the paper. Overall, a static entity has given way to one that has added to this quality a vast array of dynamic potentials. (See table 3.1.)

A common and salient feature across the whole range of information practices enacted by online newspapers in the second half of the 1990s is the uncertainty they faced about almost all the elements that constituted their enterprise, from what to produce to how to do it, and from who should do it to how to evaluate the performance of products and processes. Thus, a reporter covering a seminar on "The Print Newspaper: Its Future and Its Role" held by the American Press Institute in December 1996 began her article by stating that "the only things certain and unchanging facing the newspaper industry in the future are uncertainty and change" (Gersh Hernandez 1996, p. 9). The illustration reproduced

Table 3.1
Transformations from print to online newspapers.

An entity that has mostly been	has been extended by also being
generalized	specialized (customization)
physically bound	physically unbound (verticals)
place-bound	place-unbound (networks)
temporally bound:	temporally unbound:
• limited duration	• unlimited duration (archives)
• fixed production cycles	• variable production cycles (updates)
media bound	media unbound (multimedia)
loci-bound	loci-unbound (user-authoring)
in general, static	in general, dynamic

here as figure 3.5 (which appeared on the cover of the April 1999 issue of the trade publication *Mediainfo.com* with the legend "Do newspapers have a future on the Net?") expresses the centrality of uncertainty during this period. Molina (1997b, p. 224) also concluded that uncertainty was a key issue in his analysis of the state of the industry: "Uncertainty is the name of the game at this stage. There is uncertainty about profitability

Figure 3.5
Uncertainty about the development of online newspapers. Source: Mediainfo.com, April 1999. © Tony Champagne. Reprinted with permission.

from multimedia ventures, about what will happen to revenues from the print product."

Profitability was a particularly sensitive issue for online newspapers. A proliferation of revenue strategies, coupled with a lack of profits in the vast majority of the cases, heightened the uncertainty about the worth of any alternative.[20] A December 1997 *Editor & Publisher* article listed seven revenue streams pursued by online newspapers until then: display advertising, sponsorships, classified ads and directories, Internet access, subscriptions, transactions, and pay-per-story archives (Kirsner 1997b).[21] Despite this proliferation of income channels, the end product was an abundance of losses. For instance, approximately 90 percent of U.S. online newspapers lost money in 1996 (Levins 1997b). Two years later, at PaineWebber's annual media conference, top executives from most newspaper chains reported significant losses from their online operations, and even expected them to increase in the following year: Knight Ridder estimated losing $23 million in 1998, the Tribune Corporation $35 million, the New York Times Company between $10 million and $15 million, and Times Mirror $20 million (Neuwirth 1998b, p. 12).[22] An article in *Editor & Publisher* summarized the collective mood with regard to financial matters: "Web publishers are betting there's gold in them thar hills. The problem is, nobody knows where, how to mine it, or even for certain if it's there." (Garneau 1996, p. 2i)

However, uncertainty regarding financial matters did not slow down investment. For instance, at the end of 1999, the New York Times Company announced that it expected the losses of its web operations to grow between 100 percent and 200 percent in 2000 as a result of increased marketing and development expenses, and the president of the Washington Post Company told attendees at the last PaineWebber's media conference of the decade that, despite past losses, the company planned to spend $100 million on online ventures because "this isn't the time for neatness in Internet models" (Moses 1999a,b).

To make sense of the combination of repurposing, recombining, and recreating practices in such an uncertain context, I resort to the notion of hedging. Drawing on recent developments in economic sociology, I argue that actors' information practices constituted a form of hedging that emerged as a multidimensional response to uncertainty in a volatile operating environment.[23] This image of hedging comes not from any actor's individual and omniscient reasoning, but from the aggregate actions undertaken by the industry as a whole.

Hedging was a form of taking compensatory measures to spread risks in a volatile technical and economic context. Howard Witt, the *Chicago Tribune's* associate managing editor for interactive news, put it this way: "We're deploying across a lot of different fronts . . . because we're not sure which ones we're going to have to fight on." (Kirsner 1997a, p. 29) Because undertaking such a wide range of information practices is highly difficult for a single collective actor, online newspapers engaged in a variety of collaborative enterprises. I have shown above that several recombination and recreation initiatives were pursued by networks of collective actors composed of either newspapers from different companies, or newspapers and firms "originally" in nonmedia businesses, or newspapers and consumers, as in the case of community publishing and other forms of user authorship. This is congenial with recent studies that have suggested that "the business environment has changed in such a manner that it now rewards many of the key strengths of network forms of organization: fast access to information, flexibility, and responsiveness to changing tastes" (Powell 1990, p. 325).

Among other implications, these multi-directional strategies created an ambiguous image of the unit of production, the character of the industry, and even the identities of producers and consumers. Was a particular newspaper, CareerPath.com, or their partnership the production agent of that newspaper's online help-wanted classifieds? Were online newspapers seen as news, broadcast, telecommunication, directory, computing, or retail businesses? Were community organizations producers or consumers in the *Bergen Record* and Koz's joint self-publishing initiative? To Stark (2001, p. 78), network forms of organization "make assets of ambiguity." This is partly because, in volatile economic environments, ambiguity enables actors to move from adaptation to adaptability as a guiding operating principle. Grabher (2000, p. 6) put it this way: "Whereas the notion of adaptation implies a retrospective view, reflecting the history of responses to changing environments, adaptability looks at the future, indicating the capabilities of coping with unforeseen challenges. In fact, adaptation and adaptability are complementary concepts. Successful adaptation in the past, a perfect 'fit' with the environment, might undermine adaptability."

One illustration of this shift from adaptation to adaptability can be seen in staffing practices. Whereas in a relatively stable industry such as print newspapers, hiring is heavily dependent on skills and experience, in the nascent online news organizations, recruiting has been more premised on flexibility and learning capabilities. This has been so because in a volatile environment, specific functional competencies run the risk of becoming rapidly dated, and the more ingrained they are, the more they can hinder the development of alternative information practices. Mary Kay Blake, Director of Recruiting and Placement for Gannett Corporation's newspaper division, was quoted as saying "We're hiring more on potential and brainpower and far less on functional skills." (Stepp 1996) Owen Youngman, Director of Interactive Media for the *Chicago Tribune*, was quoted as follows: "Flexibility is really key. . . . We can't promise that what I hired you for is what I need you for 4 weeks from now." (Stone 1999, p. 31) Scott Woelfel, CNN Interactive's vice-president and editor, said that his company was looking for people willing to "learn along with all of us and sort of create the map as we go along"— not "people who are filling a mold," but "people who are sculptors of the news" (Zollman 1998, p. 21).

After more than a decade of exploring, online newspapers began establishing settlements, which led them to start hedging. This form of hedging is consistent with newspapers' culture of innovation characterized in the previous chapter. American dailies usually ran behind the development of the web, following the lead of early entrants such as Netscape and Yahoo, even though they had been experimenting with online environments before these companies were formed. However, newspapers did not stand still. In other words, they were often not first movers, but they were not immobile either. When they moved, they tended to do so in ways that either reproduced the print product, such as in cases of repurposing, or protected various elements of their print franchise from the market for classified ads to that for news. This accounts partly for the comparatively slower and less experimental paths pursued by newspapers in relation to competitors without strong ties to established media. For instance, webzines such as Feed were more imaginative when it came to new forms of journalism, user-authorship sites such as Geocities exploited people's desire for self-expression to a greater extent, and electronic commerce sites such as Monster.com were more aggressive in the online classified arena. However, reacting defensively and pragmatically sometimes led to outcomes that contradicted deeply held beliefs about the distinctiveness of print journalism. For instance, featuring constant updates ran against the notion of "news analysis" that print held as its advantage over broadcast for the second half of the twentieth century. In addition, enabling user-authored content conflicted with the long enforced separation between the spheres of consumption and production. More generally, the picture that emerges from comparing a typical print paper with its online counterpart at the end of the

1990s is one of major transformations. That discontinuous paths may spring up from a culture of continuity is the subject of this chapter's concluding section.

The Past Survives in the Future

What effects have past endeavors had on web-related initiatives? Some have argued that the fate of the larger videotex projects in the first half of the 1980s turned newspaper firms more cautious when it came to undertaking projects on the web. For instance, to Pavlik (1998, p. 168) "a number of these companies . . . lost considerable sums," and "as a result, most are taking a somewhat less risky approach this time around, investing smaller amounts in more limited trials." Although reasonable at first sight, this interpretation does not help to explain why, for instance, Knight Ridder, the largest financial loser in the videotex game and certainly the most visible one inside and outside the industry, has remained one of the most aggressive players, first in online services and then on the web.

The particular perspective on the industry's culture of innovation I have developed in the previous chapter problematizes this interpretation more generally by challenging its main assumption: that, as Miles and Thomas (1997, p. 255) put it, "videotex is *the* exemplary failure to realize expected consumer markets for [information technology]." Although this may have been so in other industries, especially those built primarily around videotex, the situation appears to be more complex in the case of newspapers. Here, it is my contention that videotex was both a failure and a success for the actors, albeit in paradoxical ways. Its immediate commercial failure was indeed a success, for it signaled that there was no "clear and present danger" to the core print business. However, this success had an element of failure, for it reduced the incentive to explore untapped territories that were then left open to new entrants, some of which became important competitors later on. This mix of success and failure loses its seemingly contradictory character in light of newspapers' culture of innovation: failures online usually meant that print was in good health—at least in the short run—but this, in turn, limited actors' ability to pursue more offensive and longer-term strategies with higher risks but potentially higher returns.

From the perspective presented above, two effects that past efforts had on the evolution of online newspapers were that a sizable portion of the industry seemed already decided that nonprint alternatives were worth exploring and that these efforts had acquainted them with basic features of the web, such as interactivity. All of which contributed to prepare a somewhat fertile ground for the industry's appropriation of the web. Although counterfactuals are always difficult to assess, it seems that without these past developments, the unfolding of web-related enterprises would have proceeded at a much slower pace than it did.

More generally, American dailies' nonprint publishing initiatives during the 1980s and the 1990s illustrate how established media deal with new technical developments that both open new horizons and challenge their ways of doing things. My account shows that actors have attempted to create a "new" entity preserving the "old" one. That is, they have tried to transform a delivery vehicle that has remained unaltered for centuries,[24] and whose permanence has anchored a complex ecology of information symbols, artifacts, and practices, while simultaneously aiming to leave the core of what they do, and are, untouched. In 1995, Arthur Sulzberger Jr., publisher of the *New York Times*, expressed this quandary during a conversation with Esther Dyson at Harvard University's Nieman Foundation new-media conference: "Our job [at the *Times*] is to take the brand we have today and to translate it for this new medium. . . . We know it's going to have to be different than what it is today. . . . Some of the parts will be shockingly familiar to all of us. Twenty and twenty-five years from now, other parts none of us can even imagine. Do I really think we need to change what it is we are? On the contrary, I think the only thing we know for sure is that we can't afford to change what we are." ("The new economics of journalism," *Nieman Reports* 49, 1995, no. 2: 38–48)

How could something simultaneously be and not be? Sulzberger's statement is so telling precisely because what appears to be a logical contradiction is rather a transparent expression of how emerging media unfold—and the usually hybrid outcomes that result from such an evolutionary process. "Early uses of technological innovations," Marvin (1988, p. 235) wrote, "are essentially conservative because their capacity to create social disequilibrium is intuitively recognized amidst declarations of progress and enthusiasm for the new. People often imagine that, like Michelangelo chipping away at the block of marble, new technologies will make the world more nearly what it was meant to be all along. . . . This is also how historical actors secure in the perception of continuity are eternally persuaded to embrace the most radical transformations. The past really does survive in the future."

How have American dailies appropriated nonprint alternatives? They have tried to change by remaining the same. More precisely, actors in the newspaper industry have been persuaded to undertake significant transformations in their pursuit of permanence. Even though innovation has been carried on reactively, defensively, and pragmatically, it has nonetheless triggered a tremendous change. Imagining a future that would be an improved, but not radically different, version of the present, newspapers have pursued innovation efforts moving them along paths divergent from those initially foreseen. Paraphrasing Marvin, one might say that perception of sameness has led to substantive difference. Hence, discontinuous events have arisen from sources of continuity. One outcome of this has been that, paraphrasing Sulzberger, contemporary online papers have been able to simultaneously be and not be: they have been able to either repurpose existing products and processes, or recombine them, or recreate them, or do everything at once.

The accounts presented in this and the previous chapter have begun to address how American dailies have approached consumer-oriented alternatives to print publishing. But, despite its value in illuminating longitudinal patterns, this approach is less suited to capture the concrete practices that mix the established repertoire of print with the novel horizons available in a digital distributed information environment. To examine some of these practices, in the next three chapters I will present in-depth case studies of initiatives by online newsrooms aimed at creating content on a regular basis and taking advantage of some of the web's distinctive potentials. Though not representative of the average situation of online newsrooms during the second half of the 1990s, these initiatives provide an adequate vantage point to look at these practices because they exhibit with great intensity some of the key processes involved in how established media appropriate new technologies from the starting point of their existing sociomaterial infrastructures.

4

Mimetic Originality: The New York Times on the Web's Technology Section

It did not happen every day but often enough to stand out as a sort of ritual. Sometime in the early afternoon we would leave the building to buy something to eat. With so many options available within walking distance—midtown Manhattan is somewhat of a culinary melting pot—it was difficult for our food preferences to coincide. We would regroup in one of the conference rooms a few minutes later. As soon as we would began to eat, someone would grab a copy of the latest issue of the *New York Post* and read the headlines aloud. Then, mixing editorial acumen and a dry sense of humor, the people sitting around the table engaged in a collective exercise of literary deconstruction. With painstaking detail and a playful attitude, they tore apart one headline after another with such a combination of surgical precision and detached passion that French poststructuralism seemed a game for amateurs.

If language, as Ferdinand de Saussure argued (1908–09), is a system of differences, the meaning of this linguistic ritual certainly emerged from a set of contrasts. To begin, the object of attention was the *Post*, but the conversation took place at the New York Times Electronic Media Company. Then, the focus was on the printed word, but the actors were part of the CyberTimes desk, a unit of the New York Times on the Web that was partly in charge of experimenting with the unique potentials of online journalism. Finally, the mood was light and playful in an environment marked otherwise by a tone of seriousness and efficiency. These crisscrossing contrasts exemplify some of the tensions this chapter addresses: remaining strongly attached to the world of print while working on the web, keeping the centrality of old journalistic repertoires while tinkering with new storytelling horizons, and being conscious of how much was at stake by virtue of being at the *Times* while running the risks involved in exploring unknown territories. As usually happens with innovation "in the wild,"[1] actors were neither paralyzed by these tensions nor did they discard them all at once with some magic black-and-white solution. On the contrary, they enacted strategies that represented various shades of gray, dealing with some issues while keeping the tension between the established ways of print and the novel possibilities of online as an ongoing background.

When the New York Times on the Web was launched, in January 1996, it featured mostly articles from the print paper. The biggest exception was CyberTimes, a new daily section that aggregated, under the same banner, all the technology stories appearing in various parts of the print *Times* with original material written primarily for the web. This was partly because, according to Rob Fixmer, the founding editor of CyberTimes, "we needed not just to learn how to translate the newspaper into HTML for readers online, but [also] how to develop new ways of reporting and new types of journalism for the online media . . . using video, audio, etc." (interview,[2] October 9, 1997) The section became a rapid success, accounting for a significant proportion of the site's traffic and increasing its visibility at a time when the vast majority of online newspapers had only repurposed material. It drew my attention, too. I discovered CyberTimes in the summer of 1996 and was immediately intrigued by the combination of a translation of print into HTML with an exploration of the unique potentials of online journalism.

In May 1998, when I entered the field, CyberTimes had become the Technology section of the Times on the Web and was publishing only original articles. These articles increasingly shared key characteristics of print journalism: most notably, the content was conveyed almost exclusively by textual means, the publication cycle was daily, the stories' length was roughly similar to those in print, and there was a dominance of one-way communication, with feedback from users and forum exchanges separated from the core news product. In addition, a growing number of CyberTimes stories were making their way onto the pages of the print paper as well as into material syndicated for use by other print outlets, and CyberTimes contributors who did not come from the ranks of the paper were being asked to write stories for print—something relatively unusual at the *Times*. It was as if an entity intended to move beyond the translation of print into HTML had become mostly the translation of HTML into print.

This chapter's seemingly oxymoronic title, Mimetic Originality, aims to capture the dynamics of weaving permanence and change in such a way that the creation of newness turned into the creative production of sameness. These dynamics resulted from interdependent communication,

Figure 4.1
The homepage of the Times on the Web's Technology section, March 5, 1998.
© The New York Times. Reprinted with permission.

technical, and organizational practices. First, the performance of editorial tasks was marked by repurposing print processes, rather than articles, into the online newsroom. In addition, interface design and media choices inscribed the user as a technically unsavvy information seeker and the producer as a traditional journalist, all of which contributed to reinforcing a relative continuity between print and online as publishing environments. The configuration of message flows also served this purpose. Although exchanges between reporters and their readers and many-to-many conversations in forums endowed the section with a more varied repertoire of information flows than print, they were somewhat compartmentalized from the traditional "we publish, you read" mindset. The ties between the online and print worlds were further reinforced by an alignment of production processes between the CyberTimes desk and its relevant counterparts at the print newsroom, which contributed to the former taking over processes prevalent at the latter.

These empirical findings point to a more general analytical issue about the construction of media: the role of technology in newsroom practices. My study suggests that the online newsroom is best understood as a sociomaterial space. Technical considerations were paramount in the work of reporters and editors. They related to how information was created, who got to participate in this process, what products resulted from it, and how the audience for these products was conceived. Furthermore, that online tools were largely used to reproduce print journalism points to an initial conclusion about the material dimension of media construction: the rejection of technological determinism, which has been a lens widely used to look at technical change in news settings.

Before addressing these findings and insights in further detail, I will reconstruct relevant aspects of the Technology section's history and organizational context.

Context and History

The New York Times Company is a publicly traded corporation which employed more than 13,000 people and generated revenues in excess of $2.8 billion during 1997 (New York Times Company 1997). Despite the company's diversified assets, the newspaper group accounted for almost 90 percent of the company's revenue that year (ibid.). With its reputation as the "paper of record" and its weekday circulation of almost 1.1 million, the print *New York Times* has been the company's flagship, both institutionally and financially.

Since the mid 1980s, several initiatives have tried to extend the franchise of the print *Times* into the electronic realm. New York Pulse, a videotex project, was launched in 1985 and folded soon afterward (Davenport 1987). TimesFax, a news digest, was delivered via facsimile to more than 150,000 subscribers in 1995. @Times was the paper's service on America Online ("*N.Y. Times* launches edition on America Online,"

Editor & Publisher, June 25, 1994, p. 117). In the mid 1990s, the leadership of the *Times* put together a multi-disciplinary team to formulate its strategy in response to the potential threat that the Internet posed to its revenue base. In 1995, Martin Nisenholtz, with a two-decade career in new media (including positions at New York University's Interactive Telecommunications Program and Ogilvy and Mather's Interactive Marketing Group), was hired to lead the electronic operation, which then took the official name New York Times Electronic Media Company. The company consisted of New York Times Business Information Services, New York Times Television, and an array of consumer online products. Starting as a very small operation, by mid 1998 it was employing a few hundred people divided into three departments: editorial, sales and marketing, and systems. Since January 1996, the company included the New York Times on the Web. The online paper operated on the fifth floor of a modern building on the Avenue of the Americas and 43rd Street, two long blocks away from the New York Times Building in Times Square. The floor featured an open architecture with a multitude of cubicles and workstations, some conference rooms, and a few offices for top managers. The newsroom occupied a separate space, connected to the other departments by a long corridor. People were younger, and their looks more informal, than what I saw on my visits to the print *Times*'s newsroom.

When I undertook my field study in mid 1998, the online paper, according to in-house statistics, had between 60 million and 80 million page views per month, and 80 percent of the traffic came from outside the New York metropolitan area. Initially the site featured mostly material repurposed from the print *Times* and other content providers such as wire services. The major exception to this dominance of "shovelware" was CyberTimes. The group that planned the Times on the Web decided to create a new online section to aggregate all the technology stories scattered among sections of the print paper and add original stories to offer "something more" to the paper's readers who would also be visiting the site. Martin Nisenholtz described CyberTimes's role as "twofold": "to provide the *Times*'s view of the daily technology world and to act as a kind of host for our experimentation with web journalism" (interview, August 28, 1998).

Rob Fixmer, from the print *Times*'s national desk, was appointed the first section editor. He believed that CyberTimes should be "the local periodical of cyberspace," by which he meant the following: "We started with the assumption that cyberspace was a real entity . . . and within [it] there existed a community that was every bit in need of information and services as a physical community would be." Thus he "decided to look at the social, economic, financial, cultural aspects the Internet" (interview, October 9, 1997). He assembled a team of freelancers to do the reporting—because the print *Times* did not want to commit fixed resources to an effort deemed as uncertain—while full-time staff undertook editing and production tasks. The content was organized into columns and stories and also included a number of "specials."[3]

The section grew rapidly during its first year. By mid 1997 it featured ten columns and an increasing number of stories, had one staff writer, and was put together by a desk that also included an editor, a deputy editor, an assistant deputy editor, a producer, and a news assistant. This growth, however, was costly. CyberTimes was then using almost one-third of the editorial department's budget but generating only about 3 percent of Times on the web traffic. As the paper's site grew and its content became more varied, the relative importance of any of its parts became comparatively smaller. In addition, technology coverage in the print paper had also been growing, especially in two sections. The business desk was allocating increasing space to topics such as e-commerce and new-media start-ups, and Circuits, a new weekly section that covered information technology from a consumer angle, was scheduled for launch in February 1998.

In November 1997, Rob Fixmer returned to the print *Times* to become technology editor at the business desk. He was replaced by John Haskins, his deputy since early 1997, who also had come from the paper's national desk. Around the same time, top officials at the Electronic Media Company assessed the progress CyberTimes had made after almost 2 years of existence. Analyses of traffic logs, surveys and focus groups, columns' content, and the department's budget led them to conclude that the section had lost some of its editorial focus and become financially too expensive, that users wanted more updates during the day, and that growth in the paper's technology coverage was increasingly limiting the territory of CyberTimes. Thus, Haskins was asked to reorganize CyberTimes. He decided to tighten and refocus the content, increase timeliness, and decrease financial expenses. The number of columns was reduced from ten to five, each covering a clear-cut subject,[4] and the financial resources devoted to the production of specials were significantly reduced. Rich Meislin, the Electronic Media Company's editor-in-chief, said: "[CyberTimes] had gotten a bit diffuse and soft. . . . I think what it is now is a tighter and better focused, but also less experimental evolution of what the *New York Times* is." (interview, August 10, 1998)

Coinciding with the launch of Circuits in February 1998, CyberTimes was renamed the Technology section of the Times on the Web. The word "CyberTimes" was retained to designate a subsection devoted to daily original online reporting. John Haskins told me that the decision was to "make a Technology umbrella that could let CyberTimes live as its original idea, which is original content. [That] would make room for Circuits, so we wouldn't have to dump Circuits into CyberTimes and have that confusion." He added this: "We could also have a place for things that were neither CyberTimes nor Circuits, which is to say primarily business technology stories from the paper." (interview, June 2, 1998) These name changes were indicative of transformations in the nature of the original reporting. Trying to find a niche between the consumer electronics and the business technology sides, CyberTimes put comparatively more emphasis on hard news, as opposed to features, and paid more attention to updates during the day. These two directions were related. As CyberTimes became more news oriented, its stories were better suited for the web as a delivery vehicle for breaking news.

During my fieldwork, the CyberTimes desk, even though the section had been renamed the unit was still called CyberTimes, consisted of five full-time employees: Haskins as editor; Susan Stellin, who had come from online technology site CNET, as deputy editor; David Gallagher, previously at business news service Bloomberg, as assistant deputy editor; Andrew Zipern, with a background in CD-ROM publishing, as producer; and Lisa Napoli, who had worked for CNN and freelanced for several print outlets, as staff writer. Haskins had an office; Stellin, Gallagher, and Zipern occupied cubicles adjacent to one another; Napoli mostly worked from home. The mood at the CyberTimes desk was about efficiency and seriousness, with little of the playfulness usually associated with new-media ventures in those days. Part of this was probably due to the organizational context. Perhaps nothing expresses this matter better than a comment I overheard while walking down the corridor one day: "We can't be the avant garde because we are the garde." That feeling of being the ultimate gatekeeper was manifested in the ritual portrayed at the beginning of this chapter in which the identity of guardians was performed unconsciously but not inconsequentially. The organizational context placed multiple demands on the gatekeepers. The CyberTimes desk was not only part of the Times's online operation but also a unit charged with an exploratory mission. Publishing in a new environment meant revisiting what had to be kept inside the gates of the news medium, and the exploration mandate led to venture outside of them.

The following three sections examine the practices that were enacted to deal with such a complex scenario.

Reporting, Editing, and Producing

For the last half-century, scholars have gone "inside the newsroom" and shed light on the interpersonal, institutional, and political dimensions of editorial work.[5] Despite these valuable contributions, research in this area has been less successful in making sense of the material dimension of news production, an issue that has become particularly pressing in view of the computerization of newsrooms in the last few decades. According to Hansen, Ward, Conners, and Neuzil (1994, p. 568), "the power of the news report to construct social and political realities is established, but the role of advanced information technologies is not well understood in the varied news settings where important reality construction takes place." Thus, the conclusion in a recent literature review (Schudson 1997, p. 147)—"hard evidence on how new technology affects the news, or even hypotheses about it, are limited"—is not surprising. As a result of accounting for the dynamics of newsroom work in this and the next two chapters, I aim to offer some general analytical insights about the use of technology in online journalism and the products that result from this process.

The CyberTimes desk dealt mostly with editing and production chores.[6] With the exception of staff writer Lisa Napoli, the reporting was mostly undertaken by two retained freelancers, a handful of columnists and relatively regular collaborators, and a larger number of more sporadic contributors.[7] Most of these people had backgrounds in print newspapers, but only some had previous expertise in the technology area, either covering it as a reporting beat or engaged with newsroom technology issues. Almost none of the regular contributors had their own office space either at the Electronic Media Company, the print Times's newsroom, or at the paper's bureaus around the world. They worked from their home offices, filed electronically, usually via email, communicated with the editors on the phone or via email, and sporadically, if ever, visited the Electronic Media Company's offices.

I interviewed seven of the most frequent contributors. When discussing their information-gathering practices, all of them said that they did not differ much from the practices they undertook when they worked for print papers.[8] One of them put it this way: "I get on the phone and talk to people, or I meet them in person [and] try to make sure that I'm spelling their names right, and that I have a faithful record of what they told me. When I come back, I try to cobble together a story that's accurate. So that's the most important part of what I do." (interview, June 27, 1998) Jeri Clausing, CyberTimes's Washington reporter, summarized the contributors' experience concerning this matter: "I work for an editor, I go out and cover stories, and it's very much the same." (interview, June 18, 1998)

One difference, though, was that contributors usually provided hyperlinks for their articles. Sometimes those links led to web pages outside the Times on the Web, most often to the organizations that acted as sources of news, such as government agencies or large corporations. These external links followed a policy, developed by Rob Fixmer during his CyberTimes editorship, of a two-step process. Links were placed within the story, but when users clicked on them, they were led to a section at the bottom of the screen that contained a disclaimer about the content of those sites and then to a direct hyperlink to the sites (figure 4.2). Fixmer did that for two reasons: "I didn't want people leaving the site before they had finished the article . . . [and] I felt that we had a real liability issue sending people to other sites . . . that we couldn't control." (interview, October 9, 1997) The potential of these "external" hyperlinks,

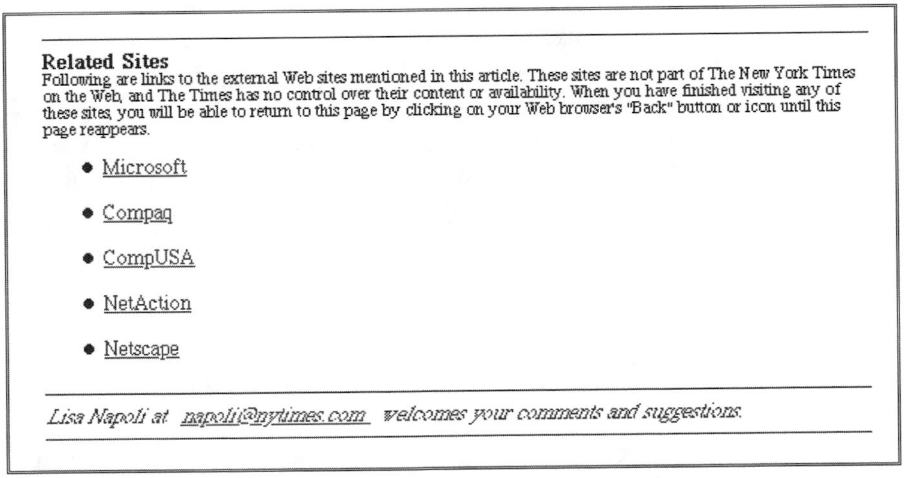

Related Sites
Following are links to the external Web sites mentioned in this article. These sites are not part of The New York Times on the Web, and The Times has no control over their content or availability. When you have finished visiting any of these sites, you will be able to return to this page by clicking on your Web browser's "Back" button or icon until this page reappears.

- Microsoft

- Compaq

- CompUSA

- NetAction

- Netscape

Lisa Napoli at napoli@nytimes.com welcomes your comments and suggestions.

Figure 4.2
The bottom of a CyberTimes article, with external hyperlinks. © The New York Times. Reprinted with permission.

as opposed to just mentioning news sources in a story, intrigued most of my interviewees since it challenged aspects of the editorial gatekeeping process central to the occupational identity of modern journalists. For instance: "How do you decide what you're going to link to, especially when almost everybody now has a web site? If I happen to make a fleeting mention of a company in my story, should I link to it?" One of the contributors to whom I talked asked that question, adding "Is that like doing PR [public relations] for the company? I'm not supposed to be doing that for them. . . . I don't know what the answers to these [questions] are, they're new questions." (interview, June 27, 1998)

Things were different in the case of hyperlinks to stories or other information, such as stock quotes, that were part of the Times on the Web. These links were usually not furnished by contributors but added by editors and producers later on, and they did not follow the two-step procedure. Their presence affected the character of storytelling by reducing the space devoted to background information within the article. This was most noticeable in the case of stories unfolding over extended periods, in which authors provided a series of links to past articles instead of including one or two paragraphs summarizing context and history.

Reporting almost always involved gathering information that could be communicated through words. Put differently, despite the web's capabilities in terms of media other than those used in print journalism, such as audio, video, computer animation, and 360° photography, articles primarily conceived for publication in CyberTimes almost always used only text. To give a better sense of the dominance of textual practices, it is worth looking at the coverage of one developing story that was prominent during my fieldwork, the anti-trust trial against the Microsoft Corporation. The CyberTimes's desk had put together a special package, called "Microsoft on Trial," that contained all the articles on this matter that had appeared in the *Times*'s print and web editions since late 1997, plus a variety of background material and a forum on the story and its implications. I looked at the use of various expressive media in the articles published in the first 7 months of this special package.[9] Two of the 87 stories from print and one of the 23 initially written for the web had audio and video in their online versions. In all cases, the audio and video content came from Associated Press TV.

With the exception of Lisa Napoli, who had a background in broadcast journalism and documentary filmmaking, most regular contributors either did not know how to use audiovisual equipment professionally or did not seem highly interested in acquiring that expertise. When I asked them about using multimedia tools for reporting purposes, I received responses such as "I'm not familiar with that" (interview, June 18, 1998), "I wouldn't know how to do it" (interview, June 17, 1998), and "frankly, for me it would be a big distraction" (ibid.).

In addition to freelancers' training and preferences, another factor that contributed to the lack of multimedia storytelling was the small amount of audio and video equipment available at the Times on the Web. Tools such as cameras, recorders, editing stations, and supplies were not visible in the online newsroom's landscape. At first glance, this could have been the newsroom of a recently built print operation. The most noticeable visual sign that this was an online newsroom were four wallpapers, located adjacent to each other on one of the walls, with taped printouts of web pages from the Times on the Web and other online news outlets. Staff members printed out pages and added written comments about them. Each of the four wallpapers was devoted to a different category: "the things we do," "the competition," "prototypes," and "the piper." It is worth noticing that most of the examples in "the competition" category were sites of national news media such as CNN, ABC News, MSNBC, and Fox, signaling the changing competitive landscape as the *Times* moved from print to the web. It is also intriguing that, although the online environment is supposedly better suited than print for building and storing this kind of evolving collective document and those producing and consuming it were working at the leading edge of "new media," they nonetheless chose to resort to "old media" resources such as printouts, handwriting, and scotch tape.

In the world of the "limitless newshole," the length of CyberTimes articles did not differ much from those coming from the print paper. The majority of articles fitted within the 700–1,200-word limit with which the desk worked, and most stories fell below the 1,000-word threshold. For the sample of stories analyzed in the coverage of the Microsoft anti-trust lawsuit, CyberTimes articles were shorter on average than those coming from the print paper and had less variation in length among them. (See table 4.1.) Why such a limit on length in the absence of newsprint's physical constraints? According to John Haskins: "Most stories can be told in 600 to 700 words. Other than that it gets into background. It's just a general idea of how much needs to be said on any one topic. . . . It's not something that's specific to the web. . . . It's something that goes back to my newspaper days." (interview, June 5, 1998) I repeatedly saw editors at the CyberTimes desk counting the number of words of the articles submitted by contributors as a simple mechanism to enforce these length preferences.

Table 4.1
Number of words in first 7 months of articles in "Microsoft on trial" package.

	Stories in print *Times*	Stories in CyberTimes
Number of stories	87	23
Average length (words)	954.11	882.52
Standard deviation (words)	387.82	279.94

Copy editing proceeded much as at many print operations, but with limited resources it was less intensive than at the print *Times* (which is famous for its elaborate copy-editing and fact-checking procedures). Contributors usually sent the material via email, either within the body of the text or as a word-processing file attached to the message, and one or more CyberTimes editors worked on the copy. Often, after a first round of copy editing, contributor and editor discussed changes to the initial version. Subsequent versions went back and forth until a satisfactory one was reached. Almost all the communication between contributors and editors involved the use of technology. In exchanges among editors, there was also a substantial use of phone and email, even when they were sitting only 6 feet apart.

Once the copy was ready, it had to be formatted before being published on the site. All the formatting was done in the "mirror" servers of the Electronic Media Company.[10] The stories repurposed from the print paper were first looked up by CyberTimes personnel in the paper's database system that stored all newsroom output. Then they were manually copied to another file, stripped of their formatting, inserted with a first round of HTML format, and transferred to the web servers available to the public. In the process of moving the files from the print paper's database to the web servers, some of the HTML code tended to get "corrupted"; thus, each file had to be opened and "cleaned." Even after formatting errors were corrected, that first round of HTML format did not necessarily include all the elements needed in a web page. For instance, if the story featured a link to related past stories, it had to be added at this stage. To do this, the web publishing tool contained a set of commands with the HTML code of a variety of formatting possibilities, such as drop cap, bio of the reporter, and internal and external links, that were accessible through a pull-down menu.[11] Thus, producers had only to look for the relevant option and insert it into the story's file. The formatting of CyberTimes's original content followed similar procedures.

The only difference was that instead of having to grab stories from the paper's database, producers added the HTML code to a text document usually sent by the contributor as an email attachment. Despite their different origins, repurposed and original stories looked alike on the web. (See figures 4.3 and 4.4.)

After formatting had been completed, every story was checked in the mirror server before being published. Since the copy had already been edited, attention focused on confirming that all the elements of each story (headline, byline, links, text, sidebars, and so on) were present, and that all links worked. Once an article was ready, its publication involved executing a command that moved the file from a mirror server to one accessible to the public. Although this seemed relatively straightforward, much like executing a "print" command in a word-processing program, the publishing tool had been built in such a way that it could not entertain two publishing requests simultaneously. That is, if a producer tried to publish a file while the system was processing another request, this second request interfered with the first one, and neither could be published correctly. To deal with this matter, producers developed a procedure that appeared somewhat surreal in such a high-tech environment. Before executing the command that moved files from the mirror to the public servers, producers yelled "Anyone publishing?" and waited a few seconds to make sure they would not get in the way of other requests. This procedure was so ingrained in the online newsroom's culture that nobody's concentration seemed to be disturbed by these sudden outbursts in the ambient noise.

Even less trivial than how to publish was the decision about when to do it. Without the limitations imposed by the economics of newsprint's production and distribution, the actual cost of the publishing process is very marginal for online newsrooms. How much of this potential for multiple daily updates was realized by the CyberTimes desk? Despite the push toward updates in the online newsroom mentioned in the preceding section, during my fieldwork the CyberTimes desk tended to publish its articles on a daily basis, toward the end of the day, even in the case of original stories filed in the early afternoon.[12] In other words, CyberTimes's stories were usually made to follow the publishing cycle of the print *Times*.[13]

What does the preceding account of editorial tasks tell us about the role of technology in online editorial work? Although there was more technical manipulation in production chores than is typical in a print newsroom, and the use of hyperlinks to external sites opened up some questions about the occupational identity of contributors, the bulk of the

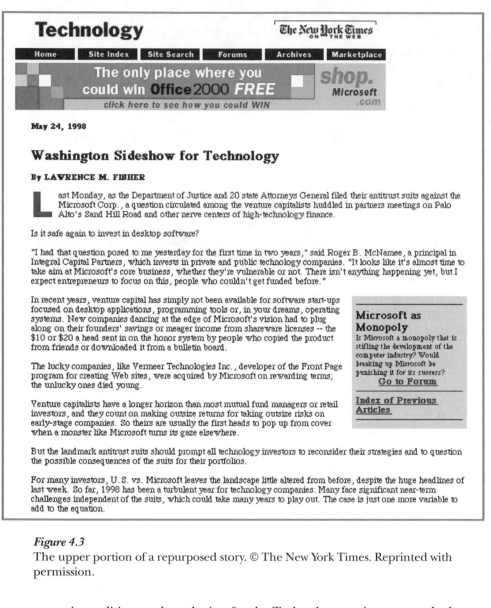

Figure 4.3
The upper portion of a repurposed story. © The New York Times. Reprinted with permission.

reporting, editing, and producing for the Technology section was marked by a repurposing pattern from print into online. But, in a paradoxical twist for a project that was born of a desire to move beyond "shovelware," actors repurposed editorial practices, rather than stories, from print. For instance, the dominance of textual material, the pre-established length (averaging fewer than 900 words), and the relatively fixed publication

January 13, 1998

At Hearing, Microsoft and U.S. Spar

By JERI CLAUSING BIO

WASHINGTON - Microsoft and the U.S. Justice Department were back in federal court on Tuesday, arguing over whether the software company is following or flouting a court order to unbundle its Internet Explorer browser from the popular Windows 95 operating system by offering inferior products.

"The government got what it wanted, knowing full well what the consequences would be," Richard J. Urowsky, a Microsoft lawyer, told U.S. District Judge Thomas Penfield Jackson.

The hearing, which continues on Wednesday, opened as reports came in from Tokyo that government officials there have searched Microsoft's Japanese offices and have also begun an investigation into whether the company's bundling of the two products is in violation of their anti-monopoly laws.

Microsoft as Monopoly
Is Microsoft a monopoly that is stifling the development of the computer industry? Would breaking up Microsoft be punishing it for its success?

Go to Forum

In Washington, the Justice Department has asked that the Redmond, Wash., company be held in contempt of Jackson's Dec. 11 order and fined $1 million a day for offering computer manufacturers just two options for separating the products: an earlier and inferior version of Windows 95 or a version of Windows that didn't work because the company said key files were omitted by removing the browser code.

Phillip Malone, a lawyer for the Justice Department, told Jackson that Microsoft was ignoring a simple option: the add/remove utility in Windows 95 that can delete programs like Internet Explorer but leaves crucial shared files that are necessary for running other programs on the operating system.

"Instead," he said, "Microsoft took extreme and illogical approach" to complying with the court. "Microsoft through its actions defied rather than complied with the order."

But Urowsky accused the government of trying to alter the court order against Microsoft.

He told the court that the government made it clear in its initial filing against Microsoft that it wanted the whole code of the Internet Explorer removed from the Windows 95 system, not just visible manifestations of the software, such as desktop icons.

Jackson questioned the relevance of Urowsky's argument, saying that the court's order and Justice Department legal papers demanded different responses from the company.

Figure 4.4
The upper portion of an original CyberTimes story. © The New York Times. Reprinted with permission.

cycle are defining traits of print newspapers. Thus, that two publishing environments—print and online—with very different technological capabilities can be associated with relatively reduced variance in editorial practices leads to a first general observation about the material dimension of online media construction: the rejection of the notion of technology-driven transformations in journalistic work. This is not to say that the features of the various technical alternatives do not matter, but rather that they do not determine by themselves the dynamics and output of newsroom practices.

Information Architecture

When CyberTimes was first designed, its "front page"[14] was similar to the cover of a magazine. (See figure 4.5.) It contained a handful of headlines, and it was updated once a day. Each front page was constructed as a single graphic file in a period of about two hours, and was saved as a graphic interchange format (GIF) file. Because it was saved as a GIF, members of the Times on the Web staff knew that it would be seen by users as they intended. However, despite advantages of this design option in terms of visual simplicity and control over user experience, editors at the CyberTimes desk and other decision makers at the Times on the Web felt that this interface presented some limitations to their journalistic practices. In August 1997 they switched to an HTML-based front page. (See figure 4.6.) Rob Fixmer wrote an open letter to CyberTimes users in which the rationale for the change was explained as follows: "That front page took more than an hour of staff time to produce, and it could offer

Figure 4.5
The old front page of the CyberTimes section, August 1, 1997. © The New York Times. Reprinted with permission.

Figure 4.6
The new front page of the CyberTimes section September 20, 1997. © The New York Times. Reprinted with permission.

readers no more than five headlines. As CyberTimes reporters began covering more breaking news . . . it became clear that to best serve our customers we needed the ability to publish our reporting more quickly. At the same time, the increasing volume of news and features we now offer translated into a need for more headline space. The old design was not able to accommodate either of those needs." (Fixmer 1997)

One thing Fixmer did not address in his letter was how much both the old look and the new one resembled the front page of the print *Times*. However, several users noted the similarities and expressed their views on a public web site that CyberTimes provided for that purpose. Two examples: "Your attempt to make the site look more like a paper newspaper . . . is anachronistic, like trying to make an automobile look like a horse and carriage." "I realize that the newspaper metaphor would be somewhat dominant in your organization. But does that mean everything has to look like a newspaper?" ("Readers respond: The new CyberTimes front page," New York Times on the Web, August 27, 1997) The resemblance to the print front page was, at least in part, a result of a deliberate decision made by Ron Louie, design director of the Electronic Media Company. The print *Times* has a six-column front page. Louie told me that he kept this scaffolding when he designed CyberTimes's front page. "In fact," he added, "the GIF section fronts were based on the six-column grid as well, [but] in a smaller version. This is the one thing I kept; it's my little . . . bridge to the paper." (interview, June 2, 1998)

This aesthetic continuity with the print *Times* was also manifest in the layout developed by the design team at the Electronic Media Company for the articles published daily on the Technology section. This layout was built using basic HTML tags supported by most of the browsers available at that time; thus, the content had a fairly traditional "document" look, with the body of the text as its main component. (See figures 4.3 and 4.4.) The layout followed from the main tenet of interface design at the Times on the Web: to design for the "lowest common technical denominator." Ron Louie put it this way: "We still design for the majority of the people. . . . That's only because people will need to read our site, [not] be entertained with all the gadgets that some other sites might have." He added that "we're purely, firstly, an information site. We have to make sure that people can get this information, at least the majority [of them]." (interview, June 2, 1998)

In the case of computer animation, this design strategy also influenced issues of media choice. As we saw in chapter 3, web designers were aware that user experience varied significantly depending on a host of technical factors, such as platform, browser, screen, connectivity tools, and so on. Thus, if the overall goal was to reach as large and diverse an audience as possible, then design practices became what Ron Louie called "a juggling act" involving a number of compromises (interview, June 2, 1998). At the time of my fieldwork, there were several animation technologies available for the web, such as Graphic Interchange Format 89 (GIF89), QuickTime VR, Flash, and Director. To describe the many differences among them is beyond the scope of this chapter. For present purposes, I will focus on one such difference. Although some options allowed for greater design possibilities but required increased bandwidth, others limited these possibilities but were easier to download. Designers at the Times on the Web tended to choose the latter, consistent with the "design for lowest common denominator" tenet. "Dealing with such a broad base of users," said one designer, "we use [animation] sparingly. The only thing we use animation for is GIF89, and that's every so often in Technology." He added that this option was chosen "because it's cross-platform. It works for everybody, and it's a nice small file size. . . . Flash and Director require a plug-in,[15] so we avoid [them] because we don't want to make our users go and get a plug-in." (interview, August 14, 1998)

These issues of interface design relate to choices concerning audio and video materials that were described in the previous sections. In interviews and informal conversations, actors expressed a variety of reasons for these choices. For instance, Susan Stellin, deputy editor of CyberTimes, emphasized product differentiation: "People go to the *Times* because they want quality editorial and reporting. If they just want a quick news clip, probably they'll go to CNN now and 10 years from now." (interview, June 11, 1998) Another member of the online newsroom emphasized representations of users' identities and preferences: "Most of our users are not interested in this yet. Usership is very low for video, because I have to download the player, and it doesn't always work. . . . I keep insisting that we do not abandon our text-only readers. After all, they are the ones who want the news; they consider any graphic a 'goodsie.'" (interview, June 2, 1998)

These processes of media choice and interface design can be, at least partly, understood by looking at the inscription of a vision of the user, the producer, and the production context in the information architecture. Recent technology scholarship has emphasized the crucial role played by constructions of the user embodied in the production of artifacts.[16] In Akrich's terms (1992, p. 208), in the very fabric of what they create, designers "inscribe" by whom and how an artifact will be used: "Designers . . . define actors with specific tastes, competences, motives, aspirations, political prejudices, and the rest, and they assume that morality, technology, science, and economy will evolve in particular ways. A large part of the work of innovators is that of 'inscribing' this vision of—or prediction about—the world in the technical content of the new object."

Although the notion of inscription has been mostly used to account for representations of the user, it can be fruitfully extended to address related representations of the producer and the production context since the three elements are intertwined.[17] First, the interface and media choices in the Technology section inscribed a notion of users who were technically unsavvy information seekers interested mostly in news content; the rest was secondary or even a "goodsie." This user inscription was tied to choices that embedded a vision of the producer as someone performing the kinds of journalistic tasks of a print reporter. Finally, the lack of use of multimedia tools, the document look of online articles, and especially, the six-column grid signaled a continuity between the Technology section and the print *Times*. If, as the old saying goes, "the devil is in the details," that every one of the hundreds of thousands pages available on the web site was designed on a scaffold similar to that of the print paper was a subtle and discrete yet powerful message that this was the online continuation of a print artifact.

This inscription of the user, the producer, and the production context was linked to a particular configuration of message flows. Much like print newspapers, most of the information available in the Technology section was communicated unidirectionally from the site to its users. In addition, actors considered this one-way flow of information composed of stories, columns, specials, and wire feeds the core of their journalistic enterprise. However, the Technology section provided a window into two information flows different from what was typical at a print newspaper. The first flow, sustained one-to-one exchanges between contributors and users, made a difference in reporting practices. The second, vibrant many-to-many communication among users in the forums, endowed the site with a conversational dimension absent in large-circulation print newspapers, and in traditional journalism more generally.

The contributors I interviewed received a certain amount of email about their articles, an option facilitated because their email addresses were usually made public at the bottom of the article. They greatly enjoyed the possibility of knowing the reactions of the users who bothered to write to them.[18] One of them put it this way: "It really keeps me on my toes. If I have a spelling mistake, I hear about it instantly from a reader. . . . More important to me, I get a lot of thoughtful comments about my reporting: different angles I could take in the future, people who agree and disagree with my take on a story, and also a lot of story ideas." (interview, June 27, 1998) The volume of email received by contributors varied depending on the article's topic. Some stories did not generate any messages; others drew more than 100 responses. In this sense, users' messages also helped contributors to have a general sense of their audience.[19] Stephen Miller, an assistant to the Technology Editor at the print *Times* and a CyberTimes columnist, said: "I don't see the market research [and] don't know what we've done in that area; I don't go to the conference rooms and ask the statistics from the web site; [hence, users' messages] sort of [give] me an idea of who's reading the column." (interview, September 25, 1998)

Studies by Massey and Levy (1999), Newhagen, Cordes, and Levy (1995), and Riley, Keough, Christiansen, Meilich, and Pierson (1998) have shown a negative attitude of journalists toward web users' feedback. However, this was not the case in my conversations with journalists either at the *Times* or at other organizations where I conducted fieldwork.[20] It was not only what my informants told me, but also a tone of enthusiasm in their voices or gestures of excitement when they spoke about this matter. In contrast, they assumed a more detached attitude when we talked about other issues. Perhaps these different findings were partly because the actors I interviewed wrote primarily for the web and were more familiar and comfortable with user feedback than print and broadcast journalists, or because they chose to work for an online publication in the first place. This welcoming of bidirectionality in information flows diverges from traditional media, where reporters usually have scant contact with, and knowledge of, their audience, and mainly use colleagues as their public. For instance, reflecting on his journalistic experience at the *New York Times* and the *Newark Star-Ledger*, Darnton (1975, p. 176) said "We really wrote for one another." In his study of newsmaking at CBS, ABC, *Newsweek*, and *Time*, Gans (1980, p. 230) found that journalists "had little knowledge about the actual audience and rejected feedback from it. . . . They filmed and wrote for their superiors and for themselves, assuming . . . that what interested them would interest the audience."

The second exception to the dominance of unidirectional message flows was the existence of a wide variety of forums.[21] During my fieldwork, the Technology section featured eleven forums in which the users posted their views on technology-related topics. Unlike the forums for the other sections of the online paper, which were managed by dedicated staff, the Technology section's forums were under the direct control of the CyberTimes desk. However, there was no fixed member of the CyberTimes desk in charge of the section's forums. In addition, neither editors nor reporters paid much attention to them. Forums were viewed as something for users, a communication space separated from their editorial activities. They almost never posted messages to the forums, and consulted them only sporadically, if at all. For instance, one regular contributor to CyberTimes replied to my question about her use of the forums by asking: "I've never been to a Times forum. . . . How does it work?" (interview, June 18, 1998)[22] Multiple reasons were given by my interviewees for this low attention. A common one was dissatisfaction with the quality of messages—for example, "I never look at the forums. . . . [I] don't have the patience for the low signal to noise ratio anymore." (interview, June 25, 1998)

What was the relationship between stories and messages in the forums? Were messages in the forums intertwined with related articles, or were they independent of the stories and columns written by the print and electronic newsrooms? To answer these questions I analyzed messages posted in the "Microsoft as Monopoly" forum—part of the "Microsoft on Trial" information package put together by the Technology section—from its launch on November 12, 1997 to June 12, 1998. During my fieldwork this was the most popular of the section's eleven forums, generating more than 4,200 posts in that seven-month period. In view of the large number of messages, I sampled three one-week periods during which important events occurred—March 2–8, April 6–12, and May 16–22.[23] These events had led the paper to publish a larger number of stories on this topic than usual, which could potentially attract increased attention of the audience and perhaps even of reporters and editors. There were 37 articles published and 814 messages posted during those three weeks. As a proxy of the relationship between the content of messages and stories, I examined whether a message contained explicit references to content other than itself and categorized those references as internal and external to the forum.[24] To see if there was any sense of conversational continuity in the forum, I further divided the internal references into two types: only one previous message and more than one previous message. To learn whether the lack of physical barriers in the distribution of newspaper information on the web had any effect on the external references, I split them into two categories: those that alluded to a story published in the *Times* and those that pointed to a news item published elsewhere. (See table 4.2.)

Several insights emerge from these results. First, the difference between internal and external references is significant: at least 60 percent of the messages referred to a previous message, whereas at most 10 percent of the messages referred to an article. Moreover, this is reinforced by another realization made while reading the messages during the coding process. In many messages the events of the story were a

Table 4.2
Percentages of internal and external references in the three samples.

	Internal references		External references	
	1 message	>1 message	*New York Times* stories	Other media stories
Sample 1	71.68	63.58	0	0.53
Sample 2	89.42	75.00	0	10.58
Sample 3	65.18	52.14	1.86	1.86

mere background to an exchange dominated by more general issues, such as antitrust regulation in the United States, the quality of Microsoft products, and the structure of the computer industry. Thus, forum participants drew on information that was both more comprehensive and less news-oriented than the content featured in most articles. Second, columns and stories published by the *Times* were referred less often than content featured in other outlets—usually other news sites—with contributors sometimes even adding hyperlinks to those articles. This raises the issue of how much loyalty exists between users of a site's forums and the site itself.[25] The relative separation between articles and messages enacted by reporters' and editors' practices was also taking place on the forum side, albeit with a reverse sign, with participants exchanging views almost regardless of the stories published on the site. Third, more than half of the messages referred to more than one message. Reading the messages it was evident that a significant part of the forum's activity was composed by several of these "conversations" running in parallel, usually including dozens of messages and lasting for several days. Sometimes these conversations got so intense that contributors responded to each other within only one minute of difference, virtually transforming an asynchronous forum into an almost synchronous chat room! This participation pattern is also evident in the distribution of contributions. During the three one-week periods, the most active quarter of participants posted approximately three-quarters of the messages. (See table 4.3.)

From reading the messages posted during the three weeks, the image that comes to mind—an image that is reinforced by the quantitative findings—is of a large number of people gathered to discuss matters related to an event in a place provided by a host whose presence remains vague and in the background. As in many large social events, most of the participants come and go fairly rapidly. However, those who stay longer tend

Table 4.3
Distribution of participation in the three samples.

	Messages	Contributors	Percentage of messages contributed by top quartile
Sample 1	173	56	71.10
Sample 2	104	24	76.92
Sample 3	537	191	73.34

to cluster in fairly stable groups and exchange views mostly among themselves. Insofar as such an image comes from reading fewer than 1,000 messages posted in only one of the more than 150 forums featured by the Times on the Web, my conclusions are far from generalizable. However, when I shared some of these findings with Justin Peacock and Cynthia Toletino, who oversaw the Times on the Web forums, they were not at all surprised. Both agreed that the image described above, although not applicable to all cases, resonated with the dynamics of many other forums that were under their management (interview, August 14, 1998).

This examination of the "Microsoft as Monopoly" forum illustrates the co-existence of diverse information flows in online publishing and the complex relationship among them. On the one hand, in contrast with the almost exclusive presence of one-to-many flows in print newspapers, the Technology section exhibited a multiplicity of information flows: uni-, bi-, and multi-directional. On the other hand, this multiplicity was somewhat compartmentalized, because the three main flows were poorly integrated, thus leading to relatively segmented information practices. This compartmentalization was even starker in the forum case, where the potential gains from extended contact among reporters, freelancers, editors, and users were not realized. In a sense, this configuration of message flows shows that the notion of the user as consumer, inscribed in interface and media choices analyzed above, was not monolithic. From another perspective, this configuration also reinforced the notion: journalists remained the dominant content providers because users' contributions were confined to areas deemed as not central by most CyberTimes actors. King (1998, p. 31) put it as follows: "Interactivity . . . does distinguish online news media from traditional ones. But redefining the relationship between news consumer and news producer will take more than the technological ability to improve two-way communication. It will take an organizational and conceptual redefinition by the media as well."

Coordinating Production

A common feature of the practices described in the previous sections is their complexity: they involve a large number of actors performing a wide array of functions and coming from diverse occupational backgrounds. Getting the job done in such a complex organizational landscape places a premium on coordinating production across boundaries, which was also prevalent in the other two settings where I conducted fieldwork. Thus, in each of the case-study chapters, I will examine cross-boundary coordination processes and resources employed by the actors. More specifically, I will focus on three boundaries identified as important in previous research on online newspapers: the boundary between the online newsroom and its print counterpart,[26] the boundary between the online newsroom and the online marketing and advertising personnel,[27] and the boundary between the online newsroom and users of the online paper.[28] Because of the co-existence of repurposed and original material and the journalistic routines and information architecture documented above, a crucial coordination locus in the case of the Technology section was the relationship between print and online newsrooms.

As could be expected when a century-old "parent" organization has to coordinate with its latest "offspring," the relationship between the print and online newsrooms had a somewhat asymmetrical quality: CyberTimes looked significantly more toward the print newsroom than vice versa. For instance, each desk at the print paper put together a list of the stories that it was working on for the next publication cycle, daily or weekly, and also for subsequent days. (See figure 4.7.) John Haskins scanned these lists into the print paper's database system many times per day to see the unfolding of technology stories coming up in the various sections of the paper, and to avoid duplication of effort. He had to actively look for that information, because the database system did not automatically forward a copy of it to the CyberTimes desks, nor did the editors at the relevant desks send him a copy by email after entering the information in the system. This process was not the most efficient way to ensure that all appropriate technology stories coming from the print paper on any day were included in the Technology section. For instance, on the afternoon of May 5, 1998, Haskins received an email from a user commenting on an op-ed piece by Thomas Friedman about the Internet in Africa that the print *Times* had published earlier that day. The initial surprised look on his face turned progressively into one of contained

```
                         CIRCUITS NOON LIST
                           May 21, 1998


POSSIBLE REFER
- - - - - - - - -        - - - - - - - - - - - - - - - - - -


PAGE A: COVER                           CLOSE TUESDAY
- - - - - - - - - - - - - - - - - - - - - - - -
NEXT Hafner JG
Microsoft's long-term plans.

BOMB Kelley JBP
People who believe they can bomb their puters with bad vibes. # With sites box
on psychokinesis, puter telepathy and suchlike? Sidebars?#


PAGE B: GEEK                            CLOSE TUESDAY
- - - - - - - - - - - - - - - - - - - - - - - -
GEEK Marriott    JG
The news.

BOOK    Lewis
Zillions of Windows 98 books on the market.


PAGE C: GAME THEORY, BRIGHT             CLOSE MONDAY
- - - - - - - - - - - - - - - - - - - - - - - - - -
GAME     Herz                     BH
Tommy Hilfiger, Snowboarding 1080, and the whole huge marketing tie-in game.

MAPS     Pams O'Connell    JBP
Sites to customize your own maps so everyone can find your house or the Westin
```

Figure 4.7
The top of Circuits' noon list for May 21, 1998. © The New York Times. Reprinted with permission.

disappointment as he read the email, as if the message reminded him of the asymmetry in the relationship between the CyberTimes desk and the print newsroom. That was the first indication that Haskins had about the existence of such a piece. Overloaded with work (the desk was two staff members short that month), he had failed to grab the story from the database system the previous day because the op-ed page did not usually carry technology columns. The user's email prompted the late inclusion of the piece in the Technology section. Although such an omission was far from being the norm, it was not the only one either.

In addition to scanning the paper's database for stories and sending CyberTimes's list to print editors, throughout the day John Haskins, and to a lesser extent his deputy and assistant editors, were in contact on the phone and via email with their peers at the relevant print desks negotiating over specific stories. Whenever more than one desk was interested in a story, the issues included what desk would have it first, who would write it, and, if a CyberTimes contributor other than its staff writer were chosen, what desk would pay for it. In general, if CyberTimes and a print desk wanted the same story, the latter would usually have precedence. That is, the story would appear in the print paper and be repurposed in the Technology section, rather than becoming a "CyberTimes extra."

Although CyberTimes's editors were very attentive to what the print paper was working on to assign original articles for the online paper, the reporters and editors who were working on technology stories at the paper were not often equally attentive to how their material could be expanded on the web. For instance, they did not suggest such additions as hyperlinks to background information, audio, video, computer animation, and forums to foster audience participation. James Gorman, editor of the Circuits section, spoke about the relationship between his desk and CyberTimes: "Mostly we concentrate on the print [product] and leave the web site to [the CyberTimes desk]." (interview, September 18, 1998) My general impression from observing work practices at the CyberTimes desk and interviewing reporters and editors at the print newsroom was of a distinction between "us and them."

This asymmetry, however, was not stable, and had decreased thanks to an increasingly intense and fluid relationship between the CyberTimes desk and the print newsroom. Two important events serve to illustrate this evolution. On the one hand, although planning for Circuits had begun in early 1996, around the same time CyberTimes debuted online, the two projects had followed separate tracks until a few months before Circuits was launched, when the editors of CyberTimes were invited to meetings to discuss story ideas and related matters. On the other hand, in mid 1998, the top editors at the print *Times* began hosting weekly meetings to deal with coordination issues arising from the increase in technology coverage, and CyberTimes's editor and deputy editor were invited as regular participants. CyberTimes Washington reporter, Jeri Clausing, told me: "I've seen . . . a much better coordination between the paper and CyberTimes . . . [which] used to be its own little entity, and it didn't use to be much communication between the technology editor at the paper and CyberTimes." Now, she said, "the paper calls me and asks me to do stories for the paper; and the paper picks up what I've done for the web site. And I coordinate now more often with the paper's reporters." (interview, June 18, 1998)

Crucial in this trend toward increased fluidity was the return of Rob Fixmer, CyberTimes's founding editor, to the print newsroom as technology editor. Fixmer brought with him a deep knowledge of the routines, needs, and resources of the CyberTimes desk. This was complemented by the fact that he was succeeded by John Haskins, his deputy editor for more than a year, which allowed the work relationship that had developed between them at the Electronic Media Company to continue in their exchanges across the print-online border. In addition to their constant negotiations over specific stories and reporters, Fixmer and Haskins also talked about general issues and long-term planning. Stephen Miller, Assistant to the Technology Editor at the print *Times* as well as a CyberTimes columnist, told me this: "We [in the print newsroom] are starting to do some cross-pollination. We sent a bunch of people over to the web site and several of them came back . . . so that makes a big difference. When you get a Rob Fixmer who comes back to [the Business desk], and he has a whole new group of writers that were only working for CyberTimes, [but] never wrote to the paper because none of the people in the paper knew them. . . . Now they're starting to get in the paper." (interview, September 25, 1998)

In an organization marked by careful and measured use of symbols, a clear signal of the trend toward increased fluidity and decreased asymmetry in the relationships between the print and online newsrooms was the growing legitimacy of CyberTimes in the print newsroom. For instance, after the relocation of Rob Fixmer from CyberTimes to the print paper, the Monday section titled Business Day: The Information Industries increasingly featured material written by CyberTimes contributors, which prompted the issue of whether they should get bylines. According to Tim Rice, the information technology editor in charge of the paper's Monday Business section, Fixmer "paved the way for letting these people and their bylines . . . in the paper. . . . That's been an issue. 'It's not a *New York Times* reporter, they don't 'get a byline.' Well, somehow . . . we've been able to use those bylined articles and nobody questions that anymore." (interview, September 18, 1998)

Looking at these different coordination processes illuminates a dimension of work rarely examined in the study of media organizations. A significant part of online editorial work is to manage related efforts of people in various units of the firm. Although this work remains invisible to readers, viewers, listeners, and users because it does not show up directly in what they read, view, listen to, or otherwise use, for instance, as the quality of copy-editing does, it is nonetheless essential to the journalistic enterprise. Such "invisible" work has been conceptualized as "articulation work."[29] According to Strauss (1988, p. 164), that is work that "refers to the specifics of putting together tasks, task sequences, task clusters—even aligning larger units such as lines of work and subprojects—in the service of work flow."

The preceding account of coordination processes at the CyberTimes desk foregrounds the articulation work undertaken by members of the desk to align their production processes and products with those of the relevant desks of the print newsroom—Business, Circuits, Science, and so on.[30] The key mechanism for such articulation of alignment appears to be the emergence of a relational space at the intersection of the CyberTimes desk and its related print desks. Intentionally or not, Fixmer's promotion upon his return to the print newsroom to a technology-related position with direct implications for the daily activities of the CyberTimes desk legitimized the online operation in relation to its print counterparts. In addition, the circulation of people and ideas came full circle: it was not just a person from the print newsroom going to the online newsroom and seeking to coordinate with his former colleagues, but also the complementary case of a person from the online newsroom going to the print newsroom and seeking to coordinate with his former colleagues. The value of such a bidirectional space was increased by the growing coverage of technology matters by the paper and its use of human and symbolic resources—freelancers, editors, sources, story ideas, and so on—first cultivated by the CyberTimes desk.

The effectiveness of this positional arrangement can be partly understood in reference to work on how brokers shape opportunities for social action by being the single point of contact between two otherwise unrelated domains.[31] As defined by Fernandez and Gould (1994, p. 1457), brokerage is "a relation in which one actor mediates the flow of resources or information between two other actors who are not directly linked." The brokerage space that opened up at the intersection of CyberTimes and the relevant desks in the print newsroom after Fixmer returned to the print paper created the conditions for an articulation of alignment between the online and print newsrooms to an extent that had not been possible before. In view of the unequal weight of the print and electronic newsrooms within the overall organization, one outcome of this alignment was the trend in the history of the CyberTimes desk toward taking over information practices that were prevalent at the paper. This occurred in consonance with repurposing print editorial practices and inscribing continuity with print in issues of interface design and media choice. However, such alignment did not proceed by eliminating the differences between print and online newsrooms: the CyberTimes desk was still different from a print newsroom in staffing, routines, product, and overall mission. Otherwise, the perception among many in the print newsroom of "us and them" would not have remained.

Concluding Remarks

The Technology section, originally intended to move beyond the translation of print into HTML, became mostly the translation of HTML into print. This chapter's title, "Mimetic Originality," is intended to evoke the dynamics whereby innovation produces emulation, inverting Westney's (1987) happy phrase. I have shown that such dynamics resulted from interdependent communication, technical and organizational practices. First, performing editorial tasks at the CyberTimes desk was marked by repurposing print processes, rather than articles, into the online newsroom. In addition, interface design and expressive media choices inscribed the user as a technically unsavvy information seeker and the producer as a traditional journalist, all of which contributed to reinforce a relative continuity between print and online as publishing environments. The configuration of message flows also served this purpose. Although many-to-many conversations in forums endowed the section with a more varied configuration than print, they were somewhat compartmentalized from the still dominant "we publish, you read" mindset. Finally, the articulation of alignment between the CyberTimes desk and its relevant counterparts at the print newsroom contributed to the former taking over practices prevalent at the latter.

Why did the creation of newness turn into the creative production of sameness in the unfolding of the Technology section? Although a more elaborate answer will emerge from a comparative analysis of the three case studies undertaken in the final chapter, here I address two rather intuitive answers to this question. A first explanation attributes this path primarily to economic factors, for instance, the cost of authoring a video report is much higher than that of a text story. This could have played a part in the decrease of specials, although the financial expense of such endeavors could have been somewhat easily offset by selling sponsorships. More important, economic factors cannot directly account for the bulk of mimetic practices. For instance, the use of pre-established length and a fixed publication cycle did not save the Electronic Media Company substantial sums of money, and even resorting to small audio clips in daily stories would not have represented significant financial investment in reporting, editing, and production.

Another answer could put forward the argument that mimetic originality resulted from the persistence of the institutionalized work patterns and symbols of the *Times*, known for its organizationally conservative dynamics. However, this could not account for the fact that things started very differently. Why not repurposing from the very beginning, in the absence of major changes in the culture of the *Times*? More important, this type of explanation would miss the fact that, in general, actors did not mindlessly reproduce a set of taken-for-granted procedures because "that was the way things were." On the contrary, many times they reflected on the what, how, and why of their practices, mindfully enacting certain options and discarding competing alternatives. Thus, rather than as a cause, institutional inertia in this case should be seen as the achievement of sameness by a situated transfer of the "old" into the "new."

As I mentioned above, studies of media construction in newsroom settings have not paid enough attention to the role of technology in this process. According to Sumpter (2000, p. 335), "media sociologies . . . have lagged the technical . . . evolution of the news worker's milieu." My analysis of the Technology section shows that this is a problematic conceptual lag because technical considerations appear to be intimately associated with how the news is told, who gets to tell it, and to what kind of public. The reduced expertise in nontextual technologies and the small amounts of multimedia equipment available in the Times on the Web newsroom were inextricably tied to editorial practices centered on the reproduction of print storytelling. Furthermore, this was coupled with online journalists who exhibited an occupational identity that resembled the one of their print counterparts, as defined partly by a traditional gatekeeping function and a disregard for user-authored content. In addition, the intended site users were, to a certain extent, also characterized in relation to technical issues: their assumed low technical expertise and high interest in news content were linked to the relatively reduced emphasis placed upon multimedia and interactive possibilities. Material matters were central in the motives and practices of the actors making the news online. Thus, overlooking the materiality of editorial work runs the risk of either missing important dynamics or misunderstanding their causes and implications.

More generally, the sparse accounts of newsroom's material culture have tended to concentrate on the perceived or anticipated effects of technological innovation on newsroom routines, paying much less attention to the processes whereby these effects may arise. In this sense, McNair (1998, p. 125) has suggested that "the form and content of journalism is crucially determined by the available technology of newsgathering, production, and dissemination." Pavlik (2000, p. 299) opened his essay "The Impact of Technology on Journalism" with the statement

"Journalism has always been shaped by technology." And Sylvie and Whiterspoon (2002, p. 35) claimed that "the telegraph, telephone, and computer . . . have changed the way people work [in the newspaper industry]." Thus, Cottle (1999, p. 24) has argued that "for researchers sensitized to processes of social construction and how these inform news manufacture and shape output, discussion of technology can perhaps all too easily slide into simplistic ideas of technological determinism." In contrast to the dominant technology effects focus, my analysis has begun to shed light on the local contingencies that shape the actual consequences that using new artifacts have in online newsrooms. It is not surprising that the moral of this story is quite the opposite from the technological determinism often associated with the focus on effects: actors enacting relatively similar information practices in relation to two very different publishing environments—print and online. This underscores the value of looking at locally contingent processes instead of technology-driven end points. In the next two chapters I will continue elaborating this matter by examining various kinds of processes that shape media construction in online newsrooms. In the final chapter, I will present some conclusions arrived from a comparative analysis of the three case studies.

Vicarious Experiences: HoustonChronicle.com's Virtual Voyager

In April 1995, Cheryl Laird, a reporter from the *Houston Chronicle*'s features desk, covered Houston's eighth annual Art Car Week. Instead of jotting down notes to be assembled later into a print story, she was outfitted with a laptop computer, a digital camera, a cellular phone, and a modem and was asked to post text and pictures almost immediately on the *Chronicle*'s recently launched site. That was the beginning of Virtual Voyager, a project that took advantage of the web's technical capabilities to foster "vicarious" experiences among its users. (See figure 5.1.) Glen Golightly, who directed the project, wrote the following in a memo:

Webster's New Collegiate Dictionary defines vicarious in the following way: a. endured or done by one person substituting for another, b. acting in place of someone else or something else, [and] c. felt or experienced as if one were taking part in the experience or the feelings of another. The entire history of modern mass communication—whether Ernie Pyle's dispatches from the front line in World War II or Neil Armstrong stepping on the moon as the world watched—is living vicariously. Through the Internet we can take it even further. . . . Virtual Voyager takes a viewer as close to being on scene as possible without actually being there. (Golightly 1996, p. 1)

During the next 3 years, Virtual Voyager became a web-only multimedia features section of HoustonChronicle.com, the aim of which was to use online technologies to convey vicarious experiences related to general-interest stories. In one "virtual voyage," a camera installed in the back seat of a car allowed the web audience to see what two journalists on a month-long trip along old Route 66 were seeing, nearly in real time. In another, text-based logs and diaries, audio and video segments, and interactive tools such as publicly available email exchanges between crew members and audience members were combined to create a multidimensional portrait of a circumnavigation of the earth on a 32-foot vessel.

At Your Fingertips

HOME ● VIRTUAL VOYAGES ● VOYAGER MAGAZINE
ON THE EDGE ● AT SEA ● SPACE CHRONICLE
E-MAIL US ● MEET THE TEAM ● HOUSTONCHRONICLE.COM

Welcome to the Virtual Voyager project. Our goal is simple
-- to entertain, inform and push the World Wide Web and
multimedia technology envelope.

-- **Glen Golightly**
Virtual Voyager supervisor

New York Journal:
Alice's parents came into town and decided they wanted to see everything! It made for an exhausting visit, and Alice writes about the adventure in this week's **journal**.

1999 CART Season -- Audio report on **Dario Franchitti**. Check out the rest of the season at **Victory Lap**.

At Sea -- Turkey. Lee and Mindi are "gunkholing" along the Turkish coast and seeing the sights.

Enjoying Vietnam
Sarah and her friends travel to Vietnam. Grab your pack and go on **Walkabout**.

VOYAGER UPDATE

The Koala Pouch:
The Houston Zoo recently welcomed three new residents, Danda-Loo, Derrilin and Yannathan. Learn all about the koalas and their way of life in the **Koala Pouch**.

The world famous Aaron

a verb! Peek inside the mind of cartoonist **Aaron Warner** and look over his shoulder as he creates strips. This week's rant -- Adventure in San Antonio!

HoustonChronicle.com

Figure 5.1
The homepage of Virtual Voyager. © HoustonChronicle.com. Reprinted with permission.

In the early days of online newspapers on the web, Virtual Voyager's creativity made it one of the most innovative new-media efforts. It accounted for approximately one-fourth of the *Houston Chronicle* site's traffic, and in 1996 it brought the *Chronicle* the Newspaper Association of America's first Digital Media Award for Best Interactive Feature. This initial success enabled the *Chronicle*'s Online Content Department[1] to allocate more resources to the initiative, to increase the frequency and complexity of voyages, and to plan a number of growth strategies. Enthusiasm ran high in February 1997, when I first visited with the Virtual Voyager team. But despite this creativity, or perhaps partly because of it, things did not unfold as planned. Virtual Voyager's relative contribution to the HoustonChronicle.com's traffic decreased as the site grew. In addition, the project was unable to attract advertisers. In February 1998, when I paid another visit, the project's future was in question. Virtual Voyager progressively became a vicarious experience in another sense: it allowed members of the *Chronicle*'s online operation to get as close as possible to having a full-fledged multimedia journalistic enterprise without becoming one.

My analysis shows that, rather than being mutually exclusive, the creative and commercial kinds of vicarious experiences became the two sides of the same innovation coin, tying together established and novel practices. First, journalistic work was marked by the blending of print, audiovisual, and information systems routines in which actors combined traditional textual techniques with the use of tools such as storyboards and video cameras and the development of computer expertise. Second, interface design, media choices, and configurations of message flows embodied a vision of the user as a technically savvy consumer of content, thus making the site less attractive to users who were either less technically adept or more interested in producing information rather than merely consuming it. Third, the coordination of productive activities between the online newsroom personnel and the graphic designers was fluid. However, it was not the case with the marketing and advertising staff. All of this contributed to constructing a media artifact that was simultaneously a creative success and a commercial failure.

Making sense of these practices generates two general analytical insights. First, it continues to further our understanding of the material dimension of online newsroom dynamics. The development of virtual voyages exhibits significant departures from print journalism: the transformation of text, audio, and video from givens into options, the development of authoring practices that mixed print, broadcast, and information systems procedures, and the challenge to the occupational identity that members of the Voyager team had acquired through experience in print journalism. In addition, the differences between Virtual Voyager and the New York Times on the Web's Technology section (analyzed in chapter 4) underscore the role of local factors in shaping technology use in the newsroom. This leads to the second major analytical theme: how to understand the products that result from these practices. The idea of vicariousness resonates with claims by Bolter and Grusin (2000) that the logic of contemporary new-media products is defined by a general technological and cultural push to erase the act of mediation through a proliferation of expressive channels intended to parallel the richness of people's experience. This logic was enacted in Virtual Voyager in relation to locally dependent practices. It is thus not surprising that the relative absence of similar kinds of practices in the making of the Technology section, analyzed in chapter 4, partly accounts for the comparatively less central influence that this logic had in its products. Thus, my analysis shows the extent to which the embodiment of such new-media logic into actual products was itself contingent on local processes. This suggests that focusing exclusively on new-media products and overlooking their production processes may lead to interpret necessity into what are contingent outcomes.

Context and History

The *Houston Chronicle* belongs to the Hearst Corporation, a privately owned media conglomerate with holdings in the newspaper, magazine, television, radio, cable, and computer businesses. In early 1998 it had either complete ownership of or considerable investments in the cable TV channels Arts and Entertainment and ESPN, the print magazines *Cosmopolitan* and *Good Housekeeping*, the King Features Syndicate, and the Internet portal Netscape. The newspaper division had twelve dailies (including, in addition to the *Chronicle*, the *Albany Times Union* and the *San Francisco Examiner*), seven weeklies, and a wire service. At the dawn of the web, top managers at Hearst Newspapers produced a white paper, titled "The Information Vision," that outlined the division's future strategy for online environments. Henry "Buzz" Wurzer, Marketing Development Manager at Hearst Newspapers and a former president of the Newspaper Association of America's New Media Federation, recalled that "essentially the concept was to say to the publisher 'you're no longer just a publisher of the newspaper in your market; you're the

CEO of the local information utility.'" (interview, July 1, 1998) Echoing the "from newspaper to information business" theme that pervaded rhetoric about nonprint initiatives during the 1980s and the 1990s, Wurzer added: "That takes us from when we were pretty much a newspaper company 5 or 6 years ago . . . to now go into the whole suite of electronic products, trying to change the culture of how our people think."

To realize this vision, Hearst management encouraged and helped the corporation's newspapers to develop web sites.[2] Within this context, the *Chronicle*—one of the largest print papers in the United States[3]—began to plan its site in 1994 and launched it in the spring of 1995. Hearst Newspapers chose the *Houston Chronicle* as a leading case, testing some innovative features there first and contributing significantly to its funding—75 percent the first year, 50 percent the second, and 25 percent the third. The site grew rapidly in both content and traffic. As of December 1997 it featured more than 100,000 pages of information, and according to an ABC Interactive audit it had more than 4 million page views that month.

HoustonChronicle.com was housed on the tenth floor of the *Chronicle*'s building in downtown Houston, in space previously used for storage. A part of the paper's electronic products division, it was under the direction of Joycelyn Marek, who added those responsibilities to her job as vice-president of marketing for the print paper. The division employed people with expertise in design, editorial, marketing, sales, and technical matters. Initially there was a strong push toward inter-occupational pollination, to the point that employees were asked to have their desks adjacent to those of colleagues in occupations other than their own. The rationale was that a new medium required a new approach to the division of labor. However, after a while, workers found it easier to relate to colleagues who shared their occupational backgrounds, and in a series of decisions top management chose to reintegrate each occupational group with its counterpart at the print paper. Thus, in 1997 the online sales and marketing employees began working with their colleagues at the print operation, going on sales calls together and preparing combined print and online packages. Some months later, the technical staff moved out of the electronic products division and became a part of the larger technical staff of the whole paper. Toward the end of this chapter I will elaborate on how these larger contextual dynamics of cross-functional collaboration played a role in shaping Virtual Voyager.

Virtual Voyager was Jim Townsend's brainchild. Townsend had had a long trajectory in the newspaper business when he became part of the small team that planned the *Houston Chronicle*'s web operation in fall 1994. He was put in charge of the site's content department in early 1995. Townsend was convinced that one of the distinctive advantages of the new communication technologies was the possibility to provide a much expanded user experience than existing print and broadcast options, allowing audience members to almost be on the scene without actually having to be there. The maiden voyage tried to bring this idea to fruition by exploiting the web's potential for constant updates, with the reporter posting new material shortly after events took place. An instant success, it became the blueprint for the voyages that followed: a staff member from the print newsroom would go out in the field, report events shortly after they took place, and be technically and editorially supported by personnel in the electronic products division.

However, after a few voyages, this model turned out to be inadequate. "Nobody was sitting there hitting their reload button to see when some new piece was going to pop up there," David Galloway, a content developer for Virtual Voyager, recalled. "Instantaneous live reporting was really being wasted because it wasn't reaching anyone until much later." (interview, March 31, 1998) This was coupled with the "discovery that we needed to do more than simple newspaper-type reporting on the web. . . . The same content that we produced [in the initial voyage] could have been put in a newspaper, magazine, a pamphlet, or any other print medium." (interview, March 31, 1998) The simultaneous solution to these two "discoveries" was a refocusing of efforts from immediacy to multimedia. Voyager began using audio in mid 1995, video about a year later, and computer animation and 360° photography in 1997. This progressively increased the technical complexity not only in editing and producing the material, but also in information-gathering practices while out in the field. Audio, video, and 360° photography required a broader and more complex array of journalistic and technical skills than those usually possessed by print reporters and editors. As voyages became technically more sophisticated, print reporters ceased to carry on information gathering tasks, and Virtual Voyager staff began going into the field.

Voyager rapidly proved popular with users. Only a year after its launch, it accounted for about one-fourth of the site's traffic. It also received substantial coverage in the trade press and accolades from industry peers, such the Newspaper Association of America award mentioned above. Partly in recognition of its success and future potential and partly as a result of an internal restructuring process, Voyager was constituted as a separate unit of the content department in October 1996. Glen Golightly, a former *Chronicle* reporter who had joined the online operation shortly after its launch, was promoted to Virtual Voyager supervisor, reporting directly to Jim Townsend. Mark Evangelista, a writer from the print *Chronicle*'s sports desk, and David Galloway, who came from the business desk of the Hearst-owned *Beaumont Enterprise*, were made full-time content developers of the unit; Valerie Prilop, an intern from Houston's University of St. Thomas, joined them a year later.[4] The unit also had the part-time assistance of a designer and a programmer. The group's status as a separate team came with the expectation of increased and more regular production:[5] three to four big projects per year, a monthly small voyage called "Voyager Magazine," and a weekly column, "On the Edge," written by Galloway.

I first visited HoustonChronicle.com, focusing on the Virtual Voyager, in February 1997. I found an upbeat mood: there were all sorts of expansion plans, from syndicating Virtual Voyager across Hearst sites to spinning it off as a separate entity, and people were very enthusiastic. During an interview (February 18, 1997), Evangelista told me: "When I worked in [the] Sports [desk] I woke up in the morning and [asked myself] 'Oh, well, what I am going to do before I go to work?' Whereas now when I wake up in the morning, I say to myself, 'It's time to conquer!'"

As time went by, voyages got journalistically and technically more complex, with more personnel and time needed before and after each trip took place. The most ambitious voyages demanded up to five people in the field, some of whom devoted several weeks full-time to their parts in the voyage. This increased the financial cost of running Virtual Voyager, which at the same time saw its share of HoustonChronicle.com's traffic progressively decline, partly because the paper's site grew significantly bigger as a result of the addition of new features. For instance, by December 1997 Voyager was the seventh most visited section of HoustonChronicle.com, out of several dozens, contributing approximately 2 percent of the traffic and consuming about 5 percent of the electronic products division's budget. Making matters more complicated, Virtual Voyager had not been able to attract advertising on its own. When I met with Evangelista and Golightly at the Newspaper Association of America's Connections conference in the summer of 1997, they were still considerably enthusiastic about Virtual Voyager but less so than before.

In early 1998, coinciding with the end of Hearst's three-year financial assistance to the *Chronicle*'s online operation, it was decided to institute a policy that specified that, after the following summer, once a new voyage was planned, the sales staff would have a month to get advertising for it. If they were successful, then the voyage would take place. If they were not, then top management would decide whether to go ahead. This policy was something new to the journalists involved, since, in a traditional print newspaper, the business side is not supposed to have such a direct effect on each new editorial project to be undertaken.

Two Voyages

Each voyage was in a sense a unique media artifact. Its construction built upon a common stock of knowledge and practices developed in the creation of earlier voyages, but also presented singular challenges to the people involved. To capture this mix of commonality and singularity, I focused heavily on two major voyages whose production I could partly observe—"At Sea" and "Asleep at the Wheel"—and compared that to information gathered about other voyages through interviews, corporate documents, and analyses of web sites.

In February 1997, Lee Gunther and Mindi Miller contacted Virtual Voyager. Gunther, a businessman, and Miller, a nurse and anthropologist, were about to embark upon a three-year circumnavigation of the earth and were trying to find media outlets for stories of their journey. The print *Chronicle*'s travel section was not interested, but someone thought that the electronic products division might be. After meeting with the sailors, only a few days before their departure, Voyager staff decided to create "At Sea," a virtual voyage following Gunther and Miller's journey that would become the biggest voyage ever in terms of size, traffic, and audience feedback. (See figure 5.2.) The sailors had already equipped their 32-foot vessel with a computer and satellite connection to communicate with friends and family by email, so Voyager staff asked them to send one or two logs and pictures per week. The site's traffic began growing rapidly, as did the number of users who wanted to communicate with the sailors by email. The voyage also attracted the attention of search engines and review sites. A few months later, Galloway became the voyage's editor and supervised a redesign that could accommodate its increasingly large and diverse content. This included a daily log (projected to include more than 1,000 entries at the end of the journey in March 2000); occasional travel stories by the sailors; three to five monthly collective answers to users' emails (time and financial limitations made individual answers prohibitive); galleries of photos, audio, and video

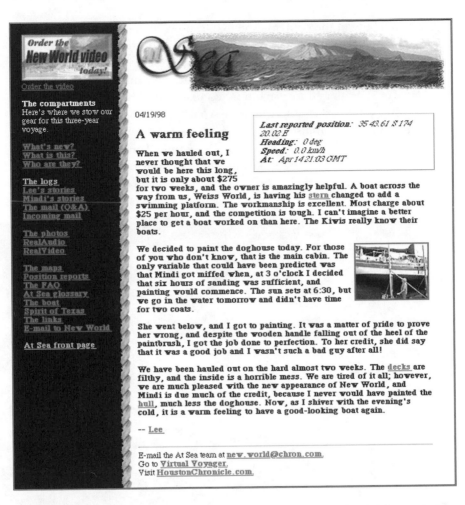

Figure 5.2
The homepage of the "At Sea" voyage. © HoustonChronicle.com. Reprinted with permission.

clips; a map that automatically updated the position of the boat twice a day; a glossary of nautical terms, with each entry hyperlinked every time it appeared in a log or a story; a Frequently Asked Questions section; a written description and photos of the boat; a collection of links to related sites on the web; and a link enabling users to send email to the sailors.

Content increased as the journey progressed. All past material was placed in a public archive in case a user discovered the site a year after its launch and wanted to learn about its beginnings. By the end of the voyage's first year, the content available to users included more than 300 logs, almost 200 pictures, 35 stories, 46 collective answers to users' emails, 26 audio reports, and 11 video compilations, among others elements.

The second voyage chronicled a tour by the western swing band Asleep at the Wheel. With a history spanning 25 years, two dozen albums, several Grammy awards, and with fame as a "road band" that would play anywhere from big stadiums to small bars, Asleep at the Wheel has been an institution of western swing music for a long time. During a voyage along Route 66 in 1996, Golightly and Evangelista met David Sanger, the band's drummer. It was from their conversations that the idea of a virtual voyage

Figure 5.3
The homepage of the "Asleep at the Wheel" voyage. © HoustonChronicle.com.
Reprinted with permission.

following the band on the road originated. The voyage took place in February 1998. It included spending four days on tour with the band and trying to construct a vicarious experience of what life on the road was like through a combination of stories, still images, audio, video, and 360° photography. Before beginning the trip, Voyager staffers worked on the addition of content and applications that would enhance that feeling of "almost being there" among its users. They developed a site with a four-part structure:

• Singin': an introduction to Asleep at the Wheel including 360° photography of the band members inside their bus, with hyperlinks to additional text and audio information about each player and a video compilation mixing original interviews with studio footage, also hyperlinked to more information using a combination of text and pictures.

• Swingin': background information on western swing including the repurposing of a piece written by Rick Mitchell, the *Houston Chronicle*'s popular music critic, and an audio interview with Mitchell enhanced with added text and pictures.

• Messin' Around: a more entertainment-focused section featuring repurposed clips of four of the band's songs and electronic postcards originally programmed and designed at the electronic products division.

• Backstage Pass: a diary of life on the road, containing mostly text information, with additional photographic, audio, and video material.

In the next sections I will look at the making of these two voyages.

Reporting, Editing, and Producing

In chapter 4 I argued that the material culture of news production has not received enough scholarly attention, and that the substantial computerization of newsrooms since the 1970s makes this gap a particularly central one in our understanding of contemporary mass media. Through an analysis of the New York Times on the Web's Technology section, I showed the extent to which novel technical possibilities can be used to reproduce editorial practices associated with prior artifacts, thus underscoring the inadequacy of the technologically deterministic narratives that characterize the popular and scholarly treatments of the role of technology in the construction of news. This chapter continues analyzing the material culture of online newsrooms through an examination of the editorial dynamics of multimedia storytelling.

One of the first changes from the world of print newspapers that I observed in the making of the Asleep at the Wheel voyage was the use of audiovisual media language, tools, and procedures. The most obvious of these was the division of the voyage into three phases: pre-production, production and post-production. I will adopt this division to structure my account.

The pre-production phase entailed generating original material, obtaining existing material, editing the material, and assembling the pieces. For instance, putting together the section that introduced the band involved scheduling a photo session with the players, obtaining the required 360° photography equipment,[6] coordinating with Mike Cowey, the *Chronicle* photographer who shot this picture, and driving to Austin, where band members lived. Once there, voyagers gathered information for biographical sketches, shot regular and 360° photographs, recorded audio interviews with each player, and a video interview with band co-founder and leader Ray Benson. Then, back in the office, all this material was assembled into the content available on the site. In the case of the band's photograph inside the bus, the 360° picture was first digitally assembled from the two halves taken by Cowey.[7] Then, the spatial parameters for each region of the picture containing a player were obtained and hyperlinked to a web page featuring that player's short biography, mug shot, and a link to an audio excerpt from the interviews. Finally, the picture's quality was digitally enhanced using graphics editing software.

Audio and video files also had to be pre-produced before they could be made available to the public, which included recording the content, editing and encoding each file using audio- and video-editing software, and moving the files to the appropriate servers. One salient issue I observed concerning pre-producing multimedia was the mixing of practices that originated in different information repertoires. First, there was a strong presence of audiovisual resources, such as the use of a storyboard that sketched the intended final clip to serve as guidance in the video-editing process, as in television production. (See figure 5.4.) Second, there was also extensive deployment of computer-related skills and practices, from knowledge of the Unix operating system to mastering video- and audio-editing software. For instance, because different connectivity tools led to somewhat different multimedia experiences, voyagers encoded each audio file for 14.4, 28.8, and ISDN (Integrated Services Digital Network), and each video file for 28.8, 56 and ISDN access.[8] Thus, when a user requested an audio file using a 33.6 modem,

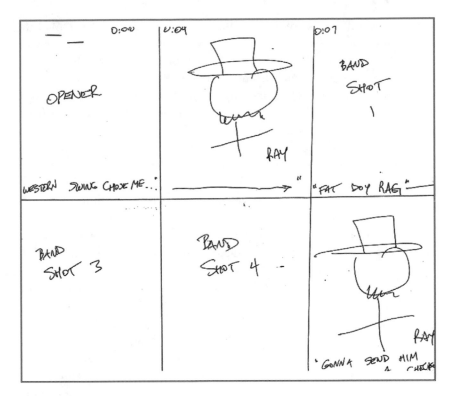

Figure 5.4
A storyboard used for video authoring. © HoustonChronicle.com. Reprinted with permission.

the *Chronicle*'s server would automatically send her the closest speed option—28.8 in this case.

Although text was used, it was an additional component of the media mix, not the privileged one. For instance, it was only in the final day of the last pre-production week that I saw a member of the Voyager team, Mark Evangelista, writing copy of significant length. When we first spoke during my February 1997 trip, Evangelista commented on the effect of less writing on journalistic practices: "You don't need to write as long because you're going to have audio [and] have video. And obviously you don't have to spend that much time describing how people look because you have video and audio. Not everyone's going to be able to do video and audio, and you're not going to be less descriptive, but you're not going to go on and on like you might do . . . if you were in print." (interview, February 18, 1997)

After several weeks' of pre-production work, Evangelista, Golightly, Prilop, and I flew to St. Louis, while Galloway stayed in Houston in charge of the "Asleep at the Wheel" voyage's backstage and other chores. The initial plan was that Evangelista and I would travel with the band in its bus for two weeks, and then be joined by Golightly and Prilop for a few days. That arrangement fell through because of last-minute opposition from the band's management: "hard fast rule: no family, no groupies, no reporters." Instead, the four of us rented a van at St. Louis's airport and followed the band, which was scheduled to play in cities hundreds of miles apart during three consecutive days: from Springfield, Missouri, we drove to Wichita, Kansas, and from there to Metropolis, Illinois.

The hours we spent driving from one venue to the next significantly cut down on the time spent on gathering material for the voyage. After arriving in town and finding a hotel, a typical day proceeded by meeting the band backstage a couple of hours before the show. By then, Evangelista, Golightly, and Prilop had already arranged what specific things they were going to focus on that particular evening. Once in the concert setting, Evangelista was in charge of doing audio interviews, jotting down notes for a concert review, and taking digital pictures; Prilop handled the video; Golightly took still pictures and coordinated all logistical matters; and I shadowed them while helping with the equipment and other low-level tasks. For the next four to six hours, voyagers gathered backstage, concert, and post-show material in a way that mixed what reporters from print and broadcast media normally do.

Things did not always go as planned, as it is often the case with the use of new technology. For instance, one of the new things that Golightly hoped to do during this voyage was to take live, as opposed to portrait, 360° photographs of concert action. In other words, the idea was to have pictures showing what the band and the audience were simultaneously doing. That could give site users a good vicarious experience of the actual concert. Taking those pictures required placing a tripod somewhere relatively equidistant from the stage and the audience. Golightly tried to do that the night the band played in Sam's Place, a dance bar on the outskirts of Wichita. What his creative impulse did not take into account, though, was that attendees were there to dance as well as to listen to the music, and they took the dance floor by storm the minute the band went on stage. Not only is technology "unruly"[9] at times; so are western swingers.

The band left town a couple of hours after the show, as soon as the roadies put all the equipment back into the bus and the truck that followed it.[10] Then, sometime past midnight, voyagers got back to their hotel rooms and worked for about three hours on the information they had gathered. Evangelista wrote a chronicle of the concert, selected, and did a quick editing of one or two digital pictures and a short audio clip to go along with the chronicle. Prilop watched the video footage and selected a 30-second segment. Golightly copy edited Evangelista's story and assisted in editing and encoding audio and video segments. As soon as everything was ready, Golightly transmitted it to a *Chronicle* server by connecting his computer to the hotel room's phone line, which, in view of the state of phone lines and the multimedia capabilities of his laptop, proved to be not a trivial matter. A couple of hours later, Galloway, in Houston, looked at the material again and posted it on the voyage's site. Then, after a few hours of sleep, we drove several hundred miles to the next venue and started all over again.

Back in Houston, Evangelista, Golightly, and Prilop worked on post-production matters, which ranged from a more careful editing of the information published while traveling, to including additional material, such as a video compilation of life on the road, to rearranging the directory structure of the whole voyage (figure 5.5). The voyage's content changed as more work was put into it, even weeks after the site went live.

If "Asleep at the Wheel" provides a window into the authoring practices of complex multimedia products, "At Sea" allows us to explore the work processes involved in recreating the genre of episodic journalism on the web. Gunther and Miller (in their very first journalistic assignments) and Galloway were mostly in charge of gathering, editing, and producing the content of "At Sea." The sailors' tasks consisted of writing logs, stories, and collective answers to emails; shooting video; taking pictures; and recording audio excerpts, which were usually taped phone calls to the electronic products division. These activities acquired a whole new character when undertaken as part of a three-year circumnavigation of the earth on a 32-foot vessel. For instance, in one of her stories, Miller wrote about having electrical problems and that one of the voyage's users suggested they avoid "stray electrical currents that could be wiping out our computer and printer motherboards. He recommended using the natural shielding of the oven to protect our equipment. You should see me sitting on the floor and typing into the oven!" (Miller 1997a).

While sailing, Miller and Gunther used their boat's satellite connection, but while in port, they located land-based connections in public places, such as libraries and cybercafés, to decrease the cost and increase the speed. They found out that fellow sailors, instead of exploring new places, were also hanging out in cyberspace. In one of the logs written

Post-voyage

Task	Assigned to	Done	Checked by	Done
Change HC.com index	Glen	XXXXXXX		
Clean up Voyager index	Glen	XXXXXXXX		
Clean up Wheel index	Vange	XXXXXXXX		
Clean up Backstage index	Vange	XXXXXXXX		
Backread -- Wichita	Glen	XXXXX		
Backread -- Springfield	Glen	XXXXXX		
Backread -- intro	Glen	XXXXXXXX		
Backread -- Metropolis	Glen	XXXXXXXXXX		
Video -- Metropolis*	Valerie	XXXXXXX		
Video -- Springfield	Valerie	XXXXXXXXX		
Video -- Wichita	Valerie	XXXXXXXXX		
Photos --Springfiled	Vange	XXXXX		
Photos -- Wichita	Vange	XXXXX		
Photos -- Metropolis	Vange	XXXXX		
Audio -- Metropolis**	Vange			
Audio -- Wichita	Vange			
Audio --Springfield	Vange			
IPIX	Glen			
Wheel history show pages	Vange			
Backread "Swing"	Glen	xxxxxxxx		
Build pages for Mitchell audio piece	Brian and Vange			
header on main index?	Brian	cancel	cancel	cancel
Video of tour	Valerie	XXXXXXXXXXXXX		
Remix mitchell.ra	Vange	XXXXXXXXXXXXX		
Check on postcards***	Glen	XXXXX		
Credits	Glen			

*Recapture and encode
**Additional encodings
*** Error message received. Is due to proxy2. Rest of world does not see.

Figure 5.5
The post-production schedule for the "Asleep at the Wheel" voyage. ©
HoustonChronicle.com. Reprinted with permission.

while stationed in New Zealand, Miller commented that she had "met several cruisers today, as usual, at the library," and that "no one seemed to be doing any reading this morning; we were all hovering around computers, trying to connect to the Internet." "Except for marinas," she added, "no other place is frequented by so many people in oil-stained T-shirts and falling-apart boat shoes!" (Miller 1998).

Mediating between sailors and the audience was Galloway's job, albeit not restricted to the more traditional editorial gatekeeping roles. He did everything from fixing the boat's communication equipment to making

"At Sea" content available to users. Galloway only had a few days to turn a system that had been configured to send and receive occasional mail to friends and family into one that could transmit larger quantities of more complex data. For instance, the boat's antenna overheated very rapidly, so it could send about 31 kilobytes of information before it automatically shut itself down to avoid melting. The digital camera given to Gunther and Miller shot photos of about 120 kilobytes each, so Galloway had to devise a way to shrink their size and turn them into coded text to reduce each picture into a size—19 kilobytes—manageable for transmission from the boat.

Shortly after leaving Houston, the sailors encountered numerous transmission problems. Galloway flew to Clearwater Beach, Florida, where the boat was docked, and configured the sailors' computers properly to handle satellite transmission. Virtual Voyager had given the sailors one of its laptops a few weeks earlier, so Galloway brought "an identical machine with me, wired them together and siphoned off every byte from their hard drive onto mine, so we have a perfect clone of a computer here for those times when technical support issues arise" (Galloway 1997d). He stayed up all night in his hotel room and by the following morning, the computers were configured correctly. Then, he went to the boat to train Gunther and Miller in data transmission, but their "efforts didn't make it to Houston. On the phone with . . . the company that provided the hardware and software for the Inmarsat-C satellite link. We ran through all the configurations and found the bugs." (ibid.) Finally, he successfully coached Gunther and Miller in how to "take a picture with the digital camera, move it to the computer, resize it, encode it and mail it back to Virtual Voyager HQ [headquarters]" (ibid.).

In addition to technical support and training, Galloway also had to come up with software solutions for the voyage's growing informational complexity. For instance, shortly after the sailors left port, users began requesting a map so they could follow their progress. After contacting vendors all over the world, Galloway discovered that although there were many options for trips within US coastal waters, no prepackaged solutions for a global journey were available. So, he posted a request for help in his weekly "On the Edge" column. It was read by none other than two programmers of the electronic products division who rapidly wrote a piece of software that captured the position of the boat twice a day and plotted it on a world map (figure 5.6).

Galloway worked at home most of the time, in a small room equipped with a desktop computer (connected to the electronic products' servers

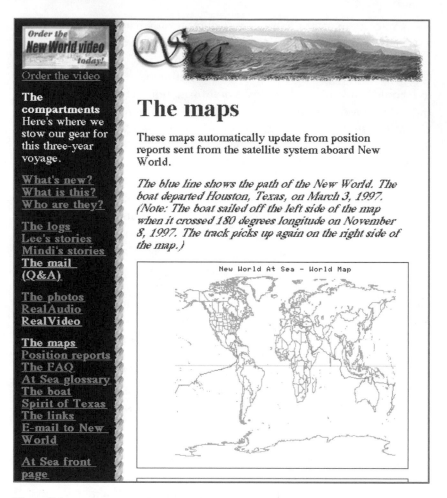

Figure 5.6
The dynamic map of the "At Sea" voyage. © HoustonChronicle.com. Reprinted with permission.

by a dedicated ISDN line) and a printer. Upon receiving the material by email, he saved it in the appropriate server and opened the file using a Unix emulator.[11] Then, he edited the copy and added the HTML format. He also inserted hyperlinks to items included in the glossary, characters, places, and situations referred to before, and other sites of relevance.[12] Then, he uploaded the new content on the server and browsed through the page, double-checking that all the elements were in place and all the hyperlinks worked. Finally, he turned all the mail sent to the sailors into a single large file, leaving only the body of the message and the sender's identification, but deleting all format and other information, and forwarded it to the boat once every few days to minimize transmission cost.

What have we learned so far about the use of the new information technologies in multimedia storytelling? In contrast to the case of the Technology section, examined in the previous chapter, newsroom practices related to Virtual Voyager exhibited a clear difference with those prevalent in a typical print setting. This difference was manifested in at least three issues: the de-reification[13] of media options, the mixing of print practices with those coming from broadcast journalism and information systems work, and the challenges all this brings to preexisting occupational identities.

To begin, some researchers have suggested that the characteristics of delivery media constrain authoring, thus affecting the choice of stories, sources, and angles.[14] For instance, Epstein (1973, p. 139) argued that high cost of audiovisual tools is related to the lower number of stories generated by television crews in comparison to print journalists: "Unlike newspapers, which generally produce more stories that can be used so that editors have room for selection, television news generally cannot afford the luxury of 'overset.'" In contrast, rather than taking the medium for granted, realizing the web's multimedia potentials moves selection processes one step earlier by requiring journalists to chose what medium or media to use for any particular story. More concretely, in creating the "Asleep at the Wheel" voyage, the team had to select what combination of text, audio, video, and 360° photography would be used for the different parts of the site. Thus, multimedia publishing de-reifies media options, turning them from givens into outcomes.

Such de-reification was tied to the acquisition of new skills. Thus, former print reporters such as Evangelista, Galloway, and Golightly had to acquire skills as far from the usual print journalist's tool kit as editing audio, shooting analog video, image mapping digital video, and storyboarding multimedia clips. Moreover, it was not just a matter of adding these new skills to the typical print storytelling abilities; voyagers also had to learn how to use text when it is only part of a more complex media ecology, which included writing about certain topics but not others, doing it more concisely, and so on.

In addition, there was an intensive technology learning process present from the very first Voyager project, when Golightly had to learn "on the fly" a set of computer abilities to technically support a reporter from the print newsroom covering a car parade in Austin. In the case of "At

Sea," Galloway's production role included an even stronger technical component. This was related to the fact that his contribution to the editorial process was not so much focused on story assignment and copy editing—what editors typically do in print newspapers—but more on supporting technically the work of two "amateur" information gatherers and facilitating their exchanges with an active set of users, some of whom also contributed directly to the site's content.

All these transformations challenged voyagers' occupational identities. For instance, during one of our interviews (February 17, 1997) Galloway said: "We all grew up in this business as writers. [But now] we're undergoing some retraining of our own thought processes to try to get away from writing long stories." Evangelista echoed a similar theme when he addressed the issue of why reporters from the print newsroom had ceased to participate in Virtual Voyager: "I don't think you can take up a regular print reporter and put him in Voyager right now. Besides the need for multimedia skills, we don't think that way, we don't think like straight print reporters now." (interview, April 3, 1998) This different way of thinking was intertwined with issues of media choice, interface design and message flows that are the subject of the next section.

Information Architecture

In the analysis of the Times on the Web's Technology section I suggested that looking at the visions of the user, the producer, and the production context that designers inscribe in the artifacts they build helps to make sense of the dynamics and the consequences of their practices. An examination of the construction of interfaces and the selection and combination of media options in the various voyages furthers our understanding of the interpenetration between the communication and technical dimensions of online journalism.

A distinctive feature of Virtual Voyager was the profusion of media elements—from text to 360° photography to video to animation—to the point that the site included a Resources page with links to the various software providers so that users could get the latest version of the many tools needed to access the material. In addition, because this happened during a period in which the web expanded dramatically, these tools were changing rapidly and continuously. The proliferation of media elements and the changing character of the tools used to create them were constant sources of conversation among voyagers and even with their audience. For instance, in an early 1997 "On the Edge" column, Galloway wrote: "Back when I started in the newspaper business a million years ago, we traveled light. A notebook, a pen and, for those really big stories, a 35mm camera. That was it. No possible need for anything else. . . . We're in a multimedia world now. Traveling to New Orleans [for a virtual voyage of Mardi Gras], I carted along about a hundred pounds of assorted electronic gear." (Galloway 1997c) He then proceeded to list the equipment, which included personal and laptop computers attached to video cameras and cell phones, digital and analog tape recorders, digital and film photograph cameras, analog video camera, and an array of supplies.

This multiplicity of media elements and tools relates to the issue of users' technical heterogeneity, first mentioned in chapter 3: that technical capabilities, skills and experience were not uniform among users, and such differences affected people's online experiences. For instance, a person downloading a video from a high-powered machine and a high-speed dedicated connection had a different visual experience from a person doing so from a three-year-old Pentium II computer and a slow modem.[15] We saw how this heterogeneity was incorporated into voyagers' production routines in the "Asleep at the Wheel" voyage, with all the audio and video files being encoded for three different access speeds. The issue also influenced the editing process. "If I were doing video for TV, I could do two-second shots and not worry about whether the audience would be able to see it," Prilop told me, but on the web she had to consider that "the average Joes at home probably are connected at 33.6 kbps, and they can't see that. I have to always be thinking about that when I'm [working], and it limits what I can shoot and use." (interview, April 1, 1998) In addition, matters of technical heterogeneity were also part of the interface design process. When Galloway managed the "At Sea" redesign process, he made sure the new interface was "backwards compatible," so that "if you have a guy with an old 486 [personal computer] and downloaded Mosaic 3 years ago and is happy running it, he can get the experience of [that voyage]." (interview, March 31, 1998)

This leads us to more general issues of interface design, another important component of Virtual Voyager's information architecture. In print newspapers, design is generally a support function of the more important editorial endeavor. In contrast, in Voyager, design played a more central role, and was seen as an integral part of the storytelling effort, which was connected to a fluid relationship between voyagers and designers that will be analyzed in the next section. This integration took place by moving away from the image of text document that dominated many newspapers sites. For instance, for the "Asleep at the Wheel" voyage, Brian Lardi, lead artist at the electronic products division, came up with a design with two salient features: a black background and a profusion of elements related to southern culture. The color was an uncommon choice in newspaper sites at the time, which tended to use light-colored backgrounds to privilege text readability. It was selected to maximize a vicarious experience of the poorly lit environment of concert venues, thus making a seamless transition between the background and the pictures and videos users would see in the foreground.

That background housed an array of elements that evoked the ambiance of western swing music, from cow horns to Texan boots. Creating each of these elements involved a laborious process. For instance, the tiny and seemingly straightforward wheel portion of the "wheel and photo icon" (figure 5.7) took more than 2 hours of a designer's work. It began by requesting carriage photos from the print *Chronicle*'s archive, choosing one that contained an appropriate wheel, and scanning it. Then, using graphics editor software, the image was turned into its negative to change the wheel's color from black to yellow; otherwise, it would not be seen against a black background. An inferior-looking part of it was restored digitally through a lengthy digital painting process. Once an image of the desired shape had been achieved, its brightness, hue, and color were edited. Although what users saw seemed

Figure 5.7
The "wheel and photo" icon. © HoustonChronicle.com. Reprinted with permission.

a flat surface, it was actually composed of many interrelated layers of elements. The design of the "wheel and photo" icon concluded by putting it into its proper place in relation to the other elements and then locking that layer into the whole frame, so that the user would download all the elements together.

How are we to understand the practices of inscribing users, producers, and the production context associated with these interface designs and media choices? A comparison with the Technology section case, analyzed in the previous chapter, helps highlight some distinctive characteristics of Virtual Voyager.

First, in contrast with the Technology section's inscription of users as technically unsavvy, Voyager people embedded an image of users as technically savvy. In Evangelista's words: "A user of a newspaper, a reader, has to pick up a newspaper and read it, [but] our users have to be Internet savvy, computer savvy, updated on operating systems and plug-ins." (interview, April 3, 1998)

Second, although interface design and media choice at the Technology section inscribed producers as professionals who were performing the kinds of journalistic tasks of print reporters, the comparable decisions at Virtual Voyager represented reporters as still performing mostly a gatekeeping role, but technically sophisticated enough to gather, process, and deliver information in as many media as seemed appropriate in each voyage. It is worth recalling that, according to Virtual Voyager staff, this was one of the factors that led to the distancing of print reporters from the production process after the first few voyages.

Third, contrary to the information architecture choices that indicated a continuity between the Technology section and the print *Times*, the inscriptions of user and producer at Voyager were linked to a production context marked by a break between print and online. Interface designs and media choices tried to make clear that Voyager was produced by a unit that was not trying to reproduce the print *Chronicle* online, or print journalism in general, but to develop something unique to the web.

These interface designs and media choices in the Voyager project were tied to a configuration of message flows that mostly had a one-way character. That is, much like print and broadcast journalism, most of the information featured on the Voyager site was communicated unidirectionally from the *Chronicle* to its users. There were forums available for users to post their views about voyages, but they were used sparingly. In addition, with a few exceptions, most voyages did not feature content contributed directly by users. Glen Golightly attributed this to the fact

that "people want to contribute to a certain point; they want to be entertained, not work" (interview, April 7, 1998).

A notable exception, however, was the very prominent role played by
"At Sea" users. It was users' positive reaction that led the Voyager team to
further develop what was initially conceived as a modest voyage. More
important for present purposes, users' desire to communicate directly
with the sailors, and Gunther's and Miller's interest in such an exchange,
triggered modifications in the site's structure to accommodate such a dialogic dimension. In an "On the Edge" column, Galloway wrote: "We're a
little overwhelmed by the response Lee and Mindi have gotten. . . . It's
just about impossible for them to give personal responses to every letter,
so the replies get lumped together under the heading of 'The Captain's
Q&A Session.'" (Galloway 1997b) As the voyage progressed, users' messages were publicly available on the site alongside sailors' responses.

But it was not only the formal aspects of this evolving site that explain
its power to accommodate a multiplicity of message flows. The character
of the exchanges also played a crucial role: "[Gunther and Miller] are
just doing something they want to do and telling it very honestly, exactly
the way it is, with all the successes and failures. . . . It's very real, and I
think a lot of people are really hooked on that." (Galloway, interview,
April 9, 1998) For instance, the boat experienced numerous problems at
the beginning of the journey, and the sailors also had difficulties performing some of the crucial tasks to keep it afloat. They wrote openly
about it, to the point that a couple of weeks after leaving Houston,
Gunther wrote in a log that he had been reflecting on "all these challenges and wondering if I am up to the task. I think I am, but if it continued day after day without letup, I doubt that I could handle the stress
of too many near misses and catastrophes" (Gunther 1997).

Gunther and Miller also shared with the audience much about their
relationship and individual feelings. This included detailed descriptions
of fights and misunderstandings that affected their sailing, as well as long
reflections on the interpersonal dynamics that arise when sharing a relatively small space with another person in the middle of the ocean. The
following fragment of one collective answer (written by Miller) to email
correspondence serves as an illustration:

To Ed D: Your remarks about our relationship were not far wrong. Lee & I weren't
even friends when we decided to attempt this circumnavigation. Then I became
seriously ill. Although Lee and his entire family showed me friendship, our beginning relationship was nevertheless stormy. New World's journey has also had its
share of bad weather—literally and figuratively. We are just now starting to form

a workable team. . . . Interestingly, we seem to view our worlds from bipolar approaches, yet we have similar personality traits. We're both stubborn; we're both sensitive and get our feelings hurt easily; we both dislike being told what to do when there's no need to dictate, etc. (Miller 1997b)

This honesty in revealing their personal, relational, and sailing journeys generated a strong sense of identification among some users. Early in the voyage, one user coined the expression "vicarious stowaway," which was then adopted by the sailors to refer to their users. "I will always remember the day when we heard from the person who coined the term 'vicarious stowaway,'" Miller told me in one of our email exchanges (May 1, 1998). "He wrote to us describing his terminal illness and he thanked us for allowing him to be a vicarious stowaway."

According to the sailors, correspondence from users often included references about how they were hooked on the voyage because Gunther and Miller were "living their dream." This feeling of having the sailors experience something on behalf of their audience, and bringing the audience on to their journey through "At Sea" was intimately tied to the conversational dynamics established between sailors and audience. These exchanges acquired an intense personal and emotional character. For instance, in one of her answers to email, Miller wrote: "We've gotten some very touching e-mail this past week from you. Your words of encouragement have been truly uplifting. . . . We're also starting to receive descriptions of your dreams—and problems—such as joblessness or alcohol abuse. Perhaps we are developing a unique bond, since sailing is a metaphor for life, and life is not always pleasant, easy or healthy." (Miller 1997c)

The intensity of these conversations among voyagers and users was not only crucial in constructing a vicarious experience for the latter, but also for keeping the former's involvement growing. "The readers are why I continue to write, dig through my soul, expose my foibles and generally try not to make [a fool] of myself that the world can see," Gunther told me. "I began asking the readers to share their lives with us: where they work, what kind of work, how many children and their interest in sailing. It was heart-warming to get their letters." (interview, May 15, 1998)

The relationship of the Voyager team, the sailors, and the vicarious stowaways provides an interesting window into an issue that has figured prominently in speculations about the potential of online journalism: that the growth of two-way and many-to-many communication in online environments may alter the character of journalism in the future.[16] For instance, Lee and So (2000, p. 13) have argued that "online journalists are expected to shift their role from information provider, gatekeeper and strong agenda-setter to 'information brokers.'" What we have seen in the present case illustrates one dimension of this type of transformation: when content emerges from ongoing conversations with multiple entry points and a very heterogeneous cast of participants, reporting is as much about "listening" as about "searching," editing is as much about "facilitating" as about "assigning" and "copy editing," and using is as much about "contributing" as about "consuming." According to Bender et al. (1996, p. 379), "the future of the industry is as much about construction as it is about consumption." In chapter 6 I will introduce another case study that took these dynamics to new heights in the world of online newspapers. For the moment I turn my attention to the coordination of production in Virtual Voyager.

Coordinating Production

In chapter 4 I argued that the complexity of the practices that create editorial products for the web highlights the importance of the coordination of key functions, occupations, and organizational units involved. In the Technology section case—in view of the co-existence of repurposed and original material, as well as the reproduction of print routines in the online newsroom—the key analytical locus was the coordination between the CyberTimes desk and its counterparts in the print paper. In contrast, creating multimedia packages at Virtual Voyager posed coordination challenges across different boundaries: those separating the occupational groups usually present in an online paper—editorial, design, systems, and marketing and advertisement. Using the approach introduced in the previous chapter, in this section I will treat the cross-boundary coordination of production in terms of "articulation work" to attempt to make visible the mechanisms and resources enacted to accomplish these processes usually "invisible to rationalized models of work" (Star 1991, p. 275).

Each occupational group participating in Virtual Voyager has its own definition of what it does, how it does it, and why other means and ends are not equally appropriate. Thus, to coordinate productive activities involving more than one of them is far from trivial. A short episode I witnessed during my fieldwork serves as a good entry to this matter. In addition to the more usual combination of text, pictures, audio, and video, the "Asleep at the Wheel" voyage also included an interactive application, electronic postcards (figure 5.8). It was conceived with the idea to promote the feeling of traveling with the band. Much like paper postcards

Figure 5.8
Electronic postcards from the "Asleep at the Wheel" voyage. ©
HoustonChronicle.com. Reprinted with permission.

are used by tourists as marks of their journey, electronic postcards were intended to enhance the vicarious experience of road life. Creating the postcards involved the joint work of a programmer and an artist. During one afternoon, while the programmer was coding in Perl,[17] the artist was creating the images with Adobe's PhotoShop, a popular graphics editing program. There was some back-and-forth between the two to coordinate the development process. After a few hours of coding, the programmer had put together a working prototype of an HTML page. This prototype enabled users to input the sender and recipient information to be processed by a CGI script,[18] which would then send the requested postcard and output another HTML page thanking the sender for having used the program. Thus, the programmer called the artist to show how the prototype, using an unattractive form as its provisional interface, worked. When he saw the page, the artist became concerned about whether its current visual manifestation would limit his creativity. The conversation turned into two parallel monologues, one focusing on functionality, and the other on visual matters, until the programmer said "I'm not a 'how it looks' person; I'm a 'how it works' person. I'm a colorblind engineer. So, tell me if this does what you need it to do, and then make it as pretty as you want!"

In a stylized version of what certainly are multifaceted occupational traits, and to paraphrase the engineer's response, editorial tends to focus on what a product "says," design on how it "looks," systems on how it "functions," and marketing and advertising on how it "sells." The case of Virtual Voyager was even more complex because, to a certain extent, each voyage presented a unique set of problems and challenges, and solutions implemented in one voyage could not always be automatically applied to a new one. Because the full-time members of the Voyager team had editorial backgrounds, and voyages were seen primarily as editorial products, in the following account I will emphasize their bilateral relationships with co-workers from design, from systems, and from marketing and advertising.

Voyager and design personnel worked well together. On the one hand, the Voyager team considered that "content" and "form" were the two sides of the multimedia storytelling coin. On the other hand, artists were eager to design the interfaces of voyages because, for them, it was an opportunity to undertake more creative projects than their usual assignments. This relationship of mutual benefit positively predisposed actors for joint work and served as a platform for mutual learning in which each group became familiar with the goals, means, and idiosyncrasies of the other. In addition, each voyage's prototypes allowed both groups to express their different viewpoints and work with a concrete object to satisfy their informational goals and needs. I attended several meetings with voyagers and designers and was struck by how easily actors from one occupation could switch frames and look at issues from the other's perspective, and by the role played by prototypes in this process. Brian Lardi, who had designed the interface of many voyages and considered himself part of the team, put it this way: "We all speak the same language. So if one day Glen [Golightly] would ask me, 'What format do you need the sound clip in, .AAIF or .WAV?' I'd say 'hey, this person speaks my language.'. . . Even though Glen comes from a more journalistic back-ground, he's like tampered into the . . . multimedia thing. We [referring to the Voyager team and the designers] all have that kind of knowledge of how things work, we know how pages are built, we know how all the pieces of the puzzle come together at the same time, and for the most part we're all on the same page." (interview, April 1, 1998)

Things did not go so smoothly between Voyager and the systems staff. According to David Galloway, some of Voyager's programming requests "were given a fairly low priority by the programming staff, and they didn't come through for us" (interview, March 31, 1998). This, in turn, "led to some strained relationships between the Virtual Voyager staff and the programming staff, and we've kind of gotten to the point where we don't even consider asking them for help if we can find any other way around it." (ibid.) The low level of cooperation between voyagers and program-mers limited the potential of some voyages. For instance, designers at the electronic products division employed Macromedia's Shockwave despite the popularity of JavaScript.[19] The advantage of JavaScript over Shock-wave was that no plug-in was required. However, Shockwave was used partly because designers felt comfortable with it since it was a continua-tion of Macromedia's Director, a tool many of them had used heavily when producing CD-ROMs in the early 1990s. But more important, nei-ther editorial nor art people involved with Voyager had the computer skills to create Java applets.[20] Mark Evangelista told me: "It would be great if we could use Java because the burden of downloading new plug-ins and updates of plug-ins drives me crazy at times. Unfortunately, we don't nec-essarily have the amount of programming necessary to develop Java applications. . . . Part of it is my fault. I should learn Java. But, as a lot of the guys will tell you, we're so busy trying to stay up with some other tech-nologies that it's hard to even think of staying up on programming appli-cations." (interview, April 3, 1998)

The relationship had not always been like that. In the early days of Voyager, there was more cooperation with programmers. However, later on, as a response to an overwhelming number of requests from both edi-torial and marketing and advertisement personnel, the electronic prod-ucts division's technical personnel formed a separate unit, centralized all the requests, and established a process whereby each request was analyzed and given a priority. That distance grew even more after the reintegration of the print and online systems personnel mentioned in an earlier section. According to Steve Newton, a media software implementer at the elec-tronic products division, "[voyagers] have an idea, they're good at finding pre-packaged things. But I think if they really wanted to kick out some-thing impressive in house, they might think a little bit more about what the programmers can do. But the downside of that, of course, is that they would have to wait in line." (interview, April 7, 1998)

Levels of priority were assigned, at least partly, according to the direct financial contribution that the final product motivating each request would make to the electronic products division. As I already noted, despite its popularity and critical acclaim, Virtual Voyager had not been able to generate much income directly, which meant that the team's program-ming requests were not always considered as promptly as voyagers wanted. This issue leads us to the relationship between voyagers and the market-ing and advertising staff. Virtual Voyager, Joycelyn Marek, the *Houston Chronicle*'s vice-president of marketing and electronic products, told me, "has been maybe one of the best publicity-learning stunts ever created because it's driven a lot of publicity our way . . . [but] it hasn't been a good revenue tool. . . . [They] were trying to get big dollars to pay for something before big dollars were there on the local level." (interview, April 2, 1998) Echoing Marek's words, others also raised the issue that although Virtual Voyager drew a significant portion of its users from outside Houston, it was nonetheless part of a site that defined itself as "Houston's leading infor-mation source." "It's been a catch 22," Jim Townsend told me. "We're stuck with the local advertising client's mindset [of] 'why do I need to be in a product that doesn't benefit me?' and from the national client who says 'but you're just Houston.'" (interview, April 1, 1998)

In addition to becoming too global a project for a site strategically ori-ented to local users, some sales representatives of the electronic products division said that Virtual Voyager was too innovative for their average potential client, who in 1997 was just starting to consider placing ads on the web. Jodie Eisenhardt, sales specialist in the electronic products divi-sion, put it this way: "[Advertisers] need to be comfortable with the ini-tial concepts of online and online newspapers. You know, the stories that are on the front page of the paper are the things that are on the front page of our site. . . . Voyager is . . . a whole different step." (interview, April 10, 1998) Also, with the trend toward integrating the print and online advertising operations and having the two staffs going on joint sales calls and offering combined buys, the sponsorship opportunities for a web-only product decreased, as did the financial incentive for the elec-tronic products division's sales representatives.

These factors—global section in a local site, a product too technically sophisticated for the average potential sponsor, and lower *de facto* incentives for selling web-only ads—were combined with the perception of a lack of adjustment between the work practices of the Virtual Voyager and sales staffs. On the one hand, some voyagers believed that sales representatives had difficulties in selling Voyager sponsorship because they did not fully understand the product. For instance, Glen Golightly told me that his "perception at times is that [members of the sales staff] don't really understand the Internet and what we are trying to do, and sometimes my feeling is that they don't want to know or spend the time to know what we're doing" (interview, April 7, 1998). On the other hand, some members of the sales staff I interviewed expressed that the way voyages were produced did not contribute to their efforts. Sometimes new voyages were decided and carried out without enough anticipation to generate a sale, whereas others were planned without much thought of their potential to attract sponsors. Cindy Hart, sales manager of the electronic products division, maintained that "in some respects it's maybe that we choose what we want to do instead of thinking what would be something we could offer an account that we have" (interview, April 10, 1998). According to Hart, that would mean "specifically going after an account and saying 'How would I sell this to Dillard's? What would Dillard's need from our department as far as a virtual voyage is concerned?' and selling that to them." (ibid.)

How are we to understand the dynamics of coordinating production in the online newspaper between the Voyager team and the other occupational groups? Comparing how things unfolded in the relationship at Voyager between the "creative" and "business" sides illuminates the role of symbolic and material resources in cross-boundary articulation work. To begin, through the course of their exchanges, voyagers and designers had generated an ability to see things from each other's perspectives and to share the "same language," as designer Brian Lardi put it. This can be seen as a process of "translation" whereby the different participants evolve common symbolic resources to communicate about the requirements of the tasks at hand.[21] Latour (1994, p. 32) says that translation "mean[s] displacement, drift, invention, mediation, the creation of a link that did not exist before and that to some degree modifies two elements or agents."

In the course of the relationship between voyagers and designers, translations unfolded in tandem with prototypes and finished products that were both outcomes and enablers of their joint work. These kinds of artifacts played a role akin to that of "boundary objects."[22] These are entities "which both inhabit several intersecting social worlds . . . *and* satisfy the informational requirements of each of them" (Star and Griesemer

1989, p. 393) In addition, boundary objects are "plastic enough to adapt to local needs and the constraints of the several parties employing them, yet robust enough to maintain a common identity across sites" (ibid., p. 393). Prototypes and finished voyages acted as boundary objects between designers and voyagers, satisfying the needs of the former for undertaking less constrained and more creative work than in their usual assignments, and of the latter for having elaborated interfaces deeply integrated with the storytelling effort—recall, for instance, the design for the "Asleep at the Wheel" voyage.

Different dynamics unfolded in the relationship between voyagers and the advertising staff. On the one hand, the former focused on developing a product that stayed on the "cutting edge" of multimedia storytelling, targeted for technically savvy and global users, which did not fit with the routines of the sales representatives. On the other hand, it was a priority for sales representatives to sell in the local marketplace combined print and web ads that focused on the average potential client, who was not very technically savvy, which was not highly congenial with the procedures of the Voyager team. Within this context, voyages satisfied the informational needs of Virtual Voyager and their users but not those of sales representatives and their potential clients. In turn, this had a negative effect on the relationship between voyagers and systems personnel. Furthermore, neither voyagers nor sales representatives were able to see the world from each other's perspective and act accordingly. Unlike the relationship between voyagers and designers, no common language originated. In the absence of shared material and symbolic resources, it is then not surprising that there was very little articulation work to coordinate the editorial and marketing and advertising actors involved.

This analysis of the dynamics of coordinating production in Virtual Voyager helps to understand its dual vicarious character introduced at the beginning of this chapter. On the one hand, it was an object that helped its users experience events as closely as possible to being on the scene without actually being there. On the other hand, it was a product that permitted the *Houston Chronicle* to experience enough of what it takes to engage in a full-fledged multimedia operation without actually having to become one. The cohesiveness between the Virtual Voyager team and the designers was critical in making voyages acquire the first meaning of vicarious experience. The lack of cohesiveness between the team and the marketing and advertisement staff was crucial in how the larger organization approached multimedia publishing in a somewhat vicarious manner.

Concluding Comments

The making of Virtual Voyager has provided a window on to some of the transformations in communication, technology, and organization involved in multimedia storytelling. To a lesser extent, as we saw in the case of users' involvement in "At Sea," it also illuminated the role of inter-activity in episodic journalism. The study suggests that editorial work was marked by a combination of practices coming from print, audiovisual, and information systems repertoires. This was tied to media choices and interface designs inscribing a discontinuity with print publishing and an exclusion of technically unsavvy users. Analyzing the various information flows enacted in Voyager has shown the dominance of traditional one-way sequences, but also the potential of more dialogic configurations. Looking at the coordination of productive activities across the different occupational groups involved has foregrounded the role of symbolic and material resources in the cohesiveness of editorial personnel with the designers and the absence of these resources in their engagement with the marketing and advertising personnel.

The unfolding of Virtual Voyager exhibited a seemingly contradictory trajectory: successful with users and industry colleagues, it nonetheless resulted in a commercial failure. These were not contradictory outcomes, but rather the two sides of the same innovation coin. The success with users and industry colleagues was mostly premised on tinkering with multimedia storytelling to an extent almost unparalleled during the early years of online papers on the web. At the same time, this created a gap between the expectations and routines of the marketing and advertising staff and the sponsors they were trying to attract. The same processes that led users to almost be on the scene without actually being there, also made corporate and advertiser constituencies experience cutting-edge multimedia journalism without fully being part of the journey.

Although counterfactuals are difficult to assess, it is not far-fetched to imagine that a tighter coupling between Voyager and the online advertising and marketing personnel would perhaps have meant better commercial prospects but less innovative storytelling. A tighter coupling between Voyager and the print newsroom, e.g., having print components of different voyages run regularly in the features section, would perhaps have increased the survival chances of the web endeavor, but maybe at the cost of decreasing its online journalism uniqueness. These hypothetical scenarios are not meant to suggest normative implications—different combinations of creative success and commercial failure present different sets of advantages and disadvantages to the actors. Rather they are meant to emphasize the extent to which the two meanings of this chapter's title, "vicarious experiences," are the dual result of a single innovation process.

Regarding more general analytical matters, the story of Virtual Voyager allows us to go deeper into the material dimension of online editorial work. Constructing voyages meant significant departures from print journalism in the practice of multimedia storytelling in at least three dimensions. First is the de-reification of media options, which means that actors had to choose whether to use text, still images, video, audio, computer animation, and 360° photography to tell either the whole story or parts of it, instead of having those choices constrained by the delivery vehicle. Second is the mixing of print routines with those coming from broadcast journalism and information systems work. For instance, voyagers intertwined textual chronicles, audiovisual material, interactive postcards, and software manipulation to produce the "Asleep at the Wheel" voyage. Third are the challenges to preexisting occupational identities, manifested, among other ways, in frequent reflections voyagers made about how their jobs compared to their print experience. The diverging paths followed by Virtual Voyager and the Technology section reinforce the conclusion reached at the end of chapter 4 about the inadequacy of technologically deterministic explanations of technological changes in online newsrooms and, in contrast, the heuristic value of looking at the local factors shaping this process.

The centrality of local contingencies points to a second set of analytical insights: how to understand the products that result from this process. An important mode of inquiry in new-media scholarship has looked at the features that distinguish recent digital artifacts from their print and audiovisual predecessors. This research has tended to focus on issues such as the elements that constitute a new artifact, the formal relationships among them, and the links between the artifact and the cultural milieu in which it originates.[23] For instance, in an illuminating rendition of this scholarship, Bolter and Grusin (2000, pp. 14–15) have proposed that new media always "remediate" their predecessors by "presenting themselves as refashioned and improved versions of other media." They have further argued that what is new about a new medium is how remediation takes place. According to them, recent digital media developments have been defined by a "double logic of remediation": "Our culture wants both to multiply its media and to erase all traces of mediation." (ibid., p. 5) For these authors, contemporary remediation "operates under the current cultural assumptions about immediacy and hypermediacy" (ibid., p. 21): "If the logic of immediacy leads one either to erase or to render automatic the acts of representation, . . . the logic of hypermediacy multiplies the signs of mediation and in this way tries to reproduce the rich sensorium of human experience." (ibid., pp. 33–34)

Immediacy and hypermediacy were central elements of Virtual Voyager. First, immediacy was manifested in the goal of creating media products that could generate a vicarious experience by bringing the user "as close to being on scene as possible without actually being there." This goal was explicitly articulated by the actors and circulated in the organizational discourse from everyday conversations to corporate memos. Second, the actors attempted to reach this goal by hypermediated means: the multiplication of the tools of representation, combining text, still images, audio, video, 360° photography, and computer animation with the intention of presenting as vivid a portrayal of events as possible. In this sense, my study of Virtual Voyager supports the claims by Bolter and Grusin (2000) about new media's double logic of remediation. However, it also departs from it, and from other new-media scholarship that focuses on products and not on their production processes, in an important way. The analysis presented in this chapter emphasizes that bringing forth remediation was a sociomaterial achievement weaving an array of practices, such as blending multiple information repertoires, inscribing users as technically unsavvy and producers as nonprint actors, and coordinating shared material and symbolic resources across the editorial and design borders. Furthermore, the absence of comparable practices in the Technology section partly accounts for the much less intense manifestation of immediacy and hypermediacy in its products, thus highlighting the contextually dependent character of these practices. This indicates that at least as important as "media logics" and "cultural wants" are the local factors shaping the appropriation of online technologies. More generally, this suggests that an exclusive or predominant focus on products without parallel attention to their actual production processes may invite reading technological or cultural determination into what are, to a great extent, context-dependent outcomes.

6

Distributed Construction: New Jersey Online's Community Connection

On September 25, 1690, in Boston, Benjamin Harris published the inaugural issue of *Publick Occurrences, Both Foreign and Domestick*, which he thought would become the first American newspaper. It had a peculiar feature: although it consisted of four pages, only three of them were printed. The fourth was left blank, according to Emery and Emery (1978, p. 22), "so the reader could add his own news items before passing it on." To this, Mott (1962, p. 10) adds "doubtless for items to be added by hand when Bostonians forwarded their papers to friends at a distance."[1] Unfortunately, the first issue of *Publick Occurrences* was also the last. The authorities banned it, since Harris did not have the necessary license required to publish a newspaper. This historical episode has long intrigued me, for it highlights to what extent the evolution of modern newspapers has followed a different trajectory. Despite the existence of features such as letters to the editor, op-ed pieces, and community pages, for the most part the creation of information has been centralized within the organization, and users have been almost entirely excluded from it. Thompson (1995, p. 29) maintains that mass communication "institutes a structured break between the production of symbolic forms and their reception. . . . [Hence] the capacity of recipients to intervene in or contribute to the process of production is strictly circumscribed."[2]

More than three centuries later, newspaper publishing on the web has given new force and meaning to user-authored content. As we saw in chapters 3–5, forums and chat rooms have been used to engage users as co-constructors of information appearing in online papers. Rankings and reviews are increasingly being opened up to the public, so that readers can post their views alongside the media "experts." But perhaps nowhere has user-authored content held more transformative potential than in "self-publishing" initiatives giving users tools to build their own web sites and host them within the online paper site. One project of this sort has been Community Connection, launched in September 1998 by New Jersey Online, the joint web site of the *Newark Star-Ledger*, the *Trenton Times*, the *Jersey Journal*, and News12 New Jersey. Community Connection enabled New Jersey nonprofit organizations to create sites within New Jersey Online where they could post information relevant to their organizations. (See figure 6.1.) Each site consisted of several sections with textual information and accompanying illustrations, and was updated as

Figure 6.1
The homepage of Community Connection. © New Jersey Online. Reprinted with permission.

often as the nonprofit wanted. The online paper did not charge non-profit organizations for publishing their sites, nor did it compensate them financially for providing content and drawing traffic to its site. After its first six months of operation, Community Connection featured more than 3,000 nonprofit sites. It also appealed to industry peers, receiving awards from *Editor & Publisher* and the Newspaper Association of America.

The making of Community Connection furnishes a privileged window to examine how actors in an online newsroom appropriate the web's interactive capabilities, an issue already explored, but to a lesser extent, in chapters 4 and 5. My study of Community Connection's first six months of operation suggests that it resulted from a new regime of information creation, which I call "distributed construction" to signal its differences from the dominant centralized mode. This new regime emerges from tying together an artifact configuration inscribing users as co-producers and enacting a multiplicity of information flows, work routines more geared to opening the gates of the site than to keeping them, and coordination resources supporting the relationships of interdependence between newsroom workers and users-turned-producers, as well as the multiple rationalities of the various groups of actors involved.

This chapter expands upon the two analytical issues raised in chapters 4 and 5. First, it adds another set of transformations related to the use of technology in online newsrooms: the shift from traditional gatekeeping to an editorial function centered on the facilitation and circulation of knowledge produced by a vast network of users-turned-producers. Second, it provides a contextual account of the practices that build inter-activity in media products, an attribute that has often been mentioned as defining new media. This ties to a third analytical dimension about which this study also yields some insights: how to understand the adoption of new-media artifacts by their public. Contrary to most work on computer-mediated communication, my research suggests that this adoption is not only dependent on the dynamics of users' online experience, but also partly contingent on the information infrastructures and user inscriptions built during the production of artifacts. Thus, making sense of how users embrace interactivity or other features of online media is severely limited if done in isolation from their mostly offline construction.

Before addressing these findings and insights in further detail, the following section introduces some aspects of the organizational environment and prior developments that influenced the unfolding of Community Connection.

Context and History

Advance Publications Incorporated is one of the largest privately owned corporations in the United States, with holdings that include print newspapers in twenty American cities, some of the most popular magazine titles in the country, such as *Vogue, Vanity Fair,* the *New Yorker,* and *Wired,* and investments in broadcast media. The corporation set up Advance Internet as a separate company to manage some of its online ventures, especially those in states where it has had print newspapers. [3]

New Jersey Online has been one of Advance Internet's leading sites since it was launched in January 1996. The site was the online presence of Advance's *Newark Star-Ledger, Trenton Times,* and *Jersey Journal* and its co-owned television station, News12 New Jersey. Its offices were located on the fifth floor of the *Jersey Journal*'s building, one floor above Advance Internet's corporate offices, overlooking Journal Square in downtown Jersey City. Similar to the Times on the Web and the HoustonChronicle.com, most people at New Jersey Online were younger than what is common in traditional media, enthusiastic about what they were doing, and happy to talk about it. According to its editor, Sara Glines, New Jersey Online was created to become "a news and information site for and about New Jersey" (interview, March 4, 1999). It rapidly gained recognition among its industry peers, receiving in its first year the Newspaper Association of America's prestigious Digital Edge Award for Best Online Newspaper. Over the years, New Jersey Online has grown into a large site. As of March 1999 it had a full-time staff of near 30 people in four departments: editorial, marketing, advertising, and production and design. According to in-house statistics, the site served more than 30 million pages in the first quarter of 1999.

Despite the differences among them, at the time of my fieldwork, all Advance Internet's newspaper sites shared two features particularly relevant to understanding Community Connection: an overall vision of the medium (often referred to as a "populist perspective") and an emphasis on local matters. Jeff Jarvis, executive vice president of Advance Internet, commented on the vision: "We're highly populist: we don't own this medium, the audience does. In all other media [it] is about us publishing to the audience, whether it's newspapers or magazines or books or television or radio. We're the gatekeepers. In this medium this is not at all true. . . . What we are really doing is enabling the audience do what they really want to do. . . . It's a very different model for publishing than any previous model." (interview, March 15, 1999)

Susan Mernit, former editorial director of New Jersey Online, explained the local emphasis as follows: "When we started we thought that the idea of local resources on the Internet was compelling because people tend to take action within 50 miles of their house." Therefore "it was very important to not only see the Internet as a global medium but as a local medium, because local is where people take action in their daily lives." (interview, April 8, 1999)[4]

Although New Jersey Online had had a "populist vision" and a local emphasis since the beginning of the site, two concrete developments triggered the creation of Community Connection. The first was the growth of stand-alone free publishing services on the web, both as models to imitate and competitors to fight. According to Jeff Jarvis, "Community Connection started because we saw the power of Geocities and Tripod."[5] (interview March 15, 1999) The second development was the success of sports forums at New Jersey Online, in terms of both number of postings and share of total site traffic. "Forums were exploding. . . . People were coming to us for information that was fairly unique, a combination of reporter- and user-generated. . . . It was potentially really cost-effective to produce because we didn't have to have a large staff to write it." (Susan Mernit, interview, April 8, 1999)

The growth of Geocities, Tripod, and other similar sites convinced Advance Internet and New Jersey Online decision makers that among their audience there could be an interest in free web publishing services and hence a potential market. Furthermore, the success of the sports, and specially soccer, forums led them to get their feet wet with a thematically restricted initiative: Youth Soccer Connection. Planning for a free web publishing initiative began in 1997 and centered around developing a prototype and a detailed set of specifications of a youth soccer-based product. Youth Soccer Connection was launched in March 1998 and, after a couple of months, it had several hundred groups publishing within it. (See figure 6.2.) Beyond its particular successes and failures, the program had a lasting effect on future initiatives aimed at engaging local users as contributors of content. According to Jeff Jarvis, "[Youth Soccer Connection] proved the concept for [New Jersey Online and Advance Internet]." (interview, March 15, 1999) The next step was extending the concept to the whole spectrum of not-for-profit activities.

Figure 6.2
The About Us/Homepage of a Youth Soccer Connection site. © New Jersey
Online. Reprinted with permission.

Information Architecture

In chapters 4 and 5 I argued that looking at the visions of the user, producer, and production context inscribed in media artifacts sheds light on
the locally contingent factors that shape their construction. An analysis of
these inscriptions in both the technical infrastructure developed to sus-
tain Community Connection and the sites built by the participating nonprofit organizations contributes to an understanding of the dynamics of

distributed construction and some of its differences with the dominant centralized mode of content creation in traditional media operations.

During 1998, New Jersey Online and Advance Internet built a publishing tool that enabled nonprofit organizations to create their web sites. Two design premises informed the tool's development. First, it should be extremely easy to use. Second, its capabilities should center on the basics of web publishing. An Advance Internet executive who was instrumental in the design process commented on the self-publishing tools existing at the time: "[They were] very confusing to me [when I] put myself in the place of a user with little computer experience. . . . My goal . . . was to make this as stupid as possible, not in a derogatory way, [so] that anybody could do this. [Because] what's important to [the potential users] was to communicate among their group and to put a face of their group to the outside world. . . . It wasn't about technology, but about communication. (interview, March 11, 1999)

Along similar lines, Susan Mernit framed the design strategy as a trade-off between usability and power. "Most software is far more powerful than the average person would ever take advantage of." When "people favor the power of a database over its usability, it becomes both very hard to use and to read. We felt it was very important to keep it simple, so that people would have success with [the publishing tool]." (interview, April 8, 1999)

These design preferences materialized in an artifact that was easy to use, requiring only that contributors have access to a computer connected to the Internet and knowledge of how to operate a keyboard and pointing device. To build their sites, users had to enter text into pre-specified spaces and make selections using a pointing device.[6] (See figure 6.3.) The resulting sites consisted of six sections with simply formatted text, one image, and up to five hyperlinks to outside sites. The design of Community Connection pages was very straightforward and practical. To achieve that simplicity and ease of use, contributors' sites were built employing almost only text formatted with basic HTML tags, such as tables and links, and some images in the form of small graphic files, to reduce download time. (See figure 6.4.) According to Jimmy Santos, one of the artists who worked in its design, the interface catered to "people who were brand new to the Internet: someone just bought a computer and didn't know what they were doing"; therefore "we had to put active words like 'click here for this', and have one big button on the front page and four icons, kind of 'you can figure it out even if you're [not an experienced user].'" (interview, March 5, 1999)

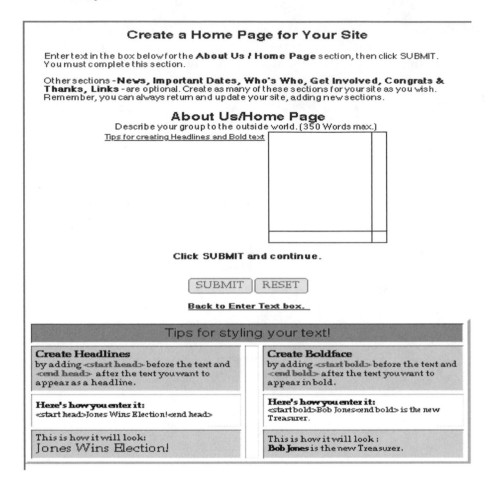

Figure 6.3
Instructions for building the About Us/Homepage of a Community Connection site. © New Jersey Online. Reprinted with permission.

What did contributors think of the publishing tool? In my interviews with representatives of nonprofit organizations that built sites within New Jersey Online, many said that they were attracted to the program partly because of the tradeoff between usability and power so carefully evaluated in designing it. The representative of a culture group said: "The tool works fairly easily. . . . The fact that one does not need to know HTML makes it a dream to use. . . . If I wanted to get fancy, I could pay someone to develop and host a site but we don't have the resources to devote to that at this time, so this works well for us." (interview, February 22, 1999)

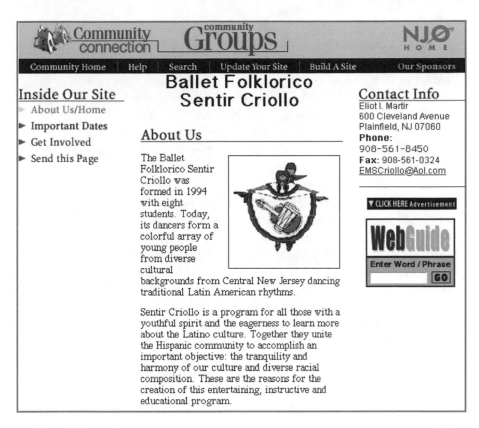

Figure 6.4
The About Us/Homepage of a community group. © New Jersey Online.
Reprinted with permission.

Sometimes I asked my interviewees whether they knew that other companies were providing free web publishing services and, if so, why they
had chosen New Jersey Online over the competing options. When they
did know about other alternatives, they almost invariably said they preferred Community Connection because of the publishing tool's ease of
use. One commented: "The only other [free web publishing service] I
looked at was Yahoo. . . . They put together a little more advanced features and formatting than what New Jersey Online people had done.
[But] you didn't really need most of the stuff that was in there; it kind of
confused the whole point of what you were trying to do. . . . That's what
I like about Community Connection: they just gave you enough to get
your ideas out there." (interview, February 11, 1999)

Although a majority of the representatives of nonprofit organizations I interviewed were content with the tool's simplicity, some expressed dissatisfaction. This dissatisfaction was apparent in the questions some contributors, who were often more technically savvy than the average contributor, asked New Jersey Online about capabilities that were not enabled by the existent version of the tool. Similar dissatisfactions surfaced in some of my interviews. Contributors said that their groups' sites would benefit from more text space, more formatting options, more graphic capabilities, multimedia options, forums, chat rooms, interactive forms, and visit counters. These requests and comments point to the "other" side of the choice of usability over power, signaling the spectrum of communication practices that were restricted by the system's design. For instance, a representative of an arts group said: "The uploading of information is idiot proof, and there lies the problem; you cannot have a user friendly web page with a lot of bells and whistles. . . . I do not like the limitations on the titles and amount of text that can be uploaded [and I] would like the ability to upload HTML pages to the site. I feel the site would be better if we could redesign our pages. They also do not give you any room to upload your own images to customize your page. . . . [Community Connection] serves our current purposes today, but [it] will not in the future if we cannot expand the layout of the web page." (interview, March 6, 1999)

In the analyses of the Times on the Web's Technology section and HoustonChronicle.com's Virtual Voyager, we saw how the people in charge of the projects inscribed in various technical choices alternative representations of the user, the producer, and the production context. In the case of Community Connection, two criteria were followed in designing the publishing tool: the preference of usability over power and the simplicity of the interface. These criteria inscribed a vision of the user as an information co-producer of community-related information. In addition, the user was inscribed as someone who should be able to take advantage of the participatory potentials of the project regardless of her or his technical capabilities and expertise, which echoes the Times on the Web's "design for lowest common denominator" strategy. As interviews with representatives of nonprofit organizations revealed, these technical choices helped attract thousands of contributors over a relatively short period.

It is worth noting that this user inscription was not hidden but was very much a part of the communication strategy adopted by New Jersey Online in its effort to enroll contributors to Community Connection. As I will describe at greater length in the next section, one of the tasks of the program's personnel was to undertake "outreach" activities (i.e., visits to schools, libraries, and community centers) to familiarize potential contributors with Community Connection or, in the actors' terms, to "evangelize the concept." One "evangelizing" occasion involved setting up a booth at the 83rd annual conference of the New Jersey State League of Municipalities, which drew thousands of attendees to Atlantic City in November 1998. I spent a day at the conference watching New Jersey Online staff members give demos of Community Connection at the booth and distribute brochures throughout the exhibit hall. I repeatedly saw them highlight the publishing tool's ease of use as one of the advantages of the program. "Anybody who can type can do it" was a phrase used time and again. This strategy seemed to resonate positively among their interlocutors, many of whom had had little if any exposure to the Internet.

A similar type of user inscription is found in other manifestations of distributed construction common in new-media sites. For instance, the software and interface chosen for user-generated rankings and reviews not only enables but also encourages users to express their opinions on a topic and share them with the audience. In contrast, readers of print newspapers are inscribed as passive individuals in terms of their information-creation capabilities. Although reading is an active process, there is nothing in the design of print newspapers as artifacts that enables, let alone encourages, readers to use them for communicating their views both to the news organization and to their fellow readers. Readers have found all sorts of non-inscribed uses for their print papers, such as wrapping fish, balancing tables, covering carpets, maintaining fires, and warming bodies. But turning them into evolving collective documents has not been achieved so far.

In Community Connection, the user inscriptions were coupled with a vision of the producer as an enabler of content creation and exchange among a vast network of contributors, which I will address more fully in the next section. This differs from the image of the producer inscribed in both traditional media as well as in the Times's Technology section and the Chronicle's Virtual Voyager. These choices regarding the publishing tool, interface design, and communication media, also inscribed a vision of the production context as a space separate from New Jersey Online's affiliated print papers in particular and the print environment in general, similar to what happened in the Voyager project.

The inscriptions of the user, the producer, and the production context were tied to particular configurations of message flows. According to

Rafaeli and LaRose (1993, p. 277), "audience-generated mass media content has invariably been subjected to a considerable degree of editorial control[7] and has generally constituted a relatively small percentage of total message system content." Therefore, it has been "more properly regarded as symbols of the community of interest rather than a true embodiment of it." I have already discussed some challenges to the typical print newspapers' configuration of message flows (chapters 4 and 5), but Community Connection is a more extreme case. To begin with, although some of the content originated at New Jersey Online (e.g., the general Community Connection homepage and those of its four components, the Frequently Asked Questions section, and the biweekly newsletter), the bulk of the information available on Community Connection came directly from the nonprofit organizations participating in the program. The same happens in other expressions of distributed construction, such as rankings and reviews of films contributed directly by users and featured alongside those written by the newspaper's critics.

Community Connection brought forth a change not only in the origin but also in the direction of messages. At one level, each of these sites constituted a one-way communication of information to its users. However, instead of information communicated from a big publisher or broadcaster to a large audience, it moved from many smaller web sites to much smaller audiences. In addition to other contact information, the availability of an email link on each site's homepage helped to increase the likelihood of two-way exchanges of information between nonprofit organizations and their audiences.[8] These transformations were going to be expanded in the "version 2.0" of the publishing tool that was being planned as a result of surveys and feedback from users. One change under consideration was to enable each contributor's site to host forums and chat rooms, thus broadening the communication potentials of the program. Jeff Jarvis put it as follows: ". . . I said [before] that all other media were about someone deciding to broadcast to the audience in one-way. Community Connection is still kind of one-way: whoever chooses to become the publisher of the site, publishes the site out. . . . [That] is not powerful enough; there's a need to allow that to become two-way." (interview, March 15, 1999) The planned addition of forum and chat capabilities could multiply the direction of information flows, from one-to-one between, for instance, a representative of a nonprofit and a user of its services, to many-to-many exchanges among members of the organization's constituency. This explosion in the direction of information flows has been a common feature of many manifestations of distributed construction, as can be seen in the many forums and chat rooms available in online newspapers.

The character of the information flows in Community Connection could be both generalized and specialized. On the one hand, the Community Connection homepage, a newsletter, or any nonprofit site were available to a potentially large audience. On the other hand, taken as a whole, Community Connection was vast and diverse enough that, in practice, the information that users accessed was only a small fraction of what was available. Moreover, this small fraction resulted from a simple customization process: the New Jersey Online server sent each user only the pages of the nonprofit sites that she or he had previously requested, with no reference to the remaining sites.[9] This duality between generalized and specialized flows also happens in other expressions of distributed construction such as forums and chat rooms, which feature content that at the same time can circulate among audiences of significant size yet in practice tends to be the concern of much smaller collectives—even to the point of including only two people communicating in a chat session.[10]

In contrast with the findings of studies that argue that online newspapers make little use of the web's interactive capabilities,[11] the evolution of Community Connection illustrates the new horizon opened by turning users into co-producers. Intimately tied to representations of producer and user embodied in the design of the publishing tool, as well as in issues of interface and media choice, a multiplicity of message flows emerged as a result of the engagement of a large and heterogeneous set of active content creators. How do these technical matters of interface design, media choice, and configuration of message flows relate to newsroom dynamics? This is important because "the very definitions of 'news' and 'newsgathering' may have to be rethought in light of new media as professional standards commingle with what might be described as amateur and grassroots reporting initiatives" (Kawamoto 1998, p. 186).

Newsroom Dynamics

The study of the Times on the Web's Technology section and HoustonChronicle.com's Virtual Voyager showed that the online newsroom is a sociomaterial space where technical considerations are central to decisions regarding what kind of stories are told, who gets to tell them, and to what kind of audience they are addressed. Moreover, I argued that new technical capabilities do not transform established journalistic routines by themselves, but that whether this takes place is an outcome of contextually dependent practices. A closer look at the work of the actors directly responsible for Community Connection illuminates the kinds of practices that bring forth a mode of information production that is significantly different from that based on traditional gatekeeping routines. In the preceding two chapters, the section that focused on journalistic work was titled Reporting, Editing, and Producing. In this chapter the corresponding section has a different title. There are two reasons for this: the personnel directly responsible for Community Connection neither reported on current events nor edited pieces undertaken by others, and their tasks were only marginally related to producing pieces for public dissemination.

Community Connection was launched in September 1998. It consisted of four sections: Community Groups, Schools, Sports, and Youth Soccer. The first three were designed from scratch; the last was carried over from Youth Soccer Connection. At the time of Community Connection's launch, there were more than 500 sites within Youth Soccer Connection as a result of almost 6 months of operation. Community Connection grew at a much faster rate; there were 1,000 sites by the end of September, 2,000 by December, and 3,000 by March 1999. Although the increase in number of sites slowed down during the second trimester of operation, my interviewees confirmed that the rate of growth in terms of traffic to these sites was increasing.[12] The program also attracted the attention of industry peers: in February 1999 it won the *Editor & Publisher*'s EPpy Award for Best Community Publishing Effort in a newspaper's online service. People at New Jersey Online and Advance Internet were happily surprised by the performance of Community Connection during its first months of existence. My perception was that the large number of sites that had been built in just a few months was what had surprised them, not that the initiative had been well received by the nonprofit sector. As I have mentioned, the early success of Youth Soccer Connection had convinced decision makers at both organizations that enabling users to author content was a sensible strategy.

New Jersey Online created two new positions within the editorial department to take primary responsibility for Community Connection. Carla Alford was hired as community producer and Betsy Old as director of community relations. Alford, who had previously spent some time at New Jersey Online as a Newspaper Association of America Diversity Fellow, devoted a considerable portion of her schedule to maintaining Community Connection's database, approving new sites, deleting incomplete ones, and, if necessary, introducing modifications to existing sites.

Alford accessed the newest submissions using the Community Connection administration tool, developed with the publishing tool.[13] She previewed the content of each new site, making sure that it was compliant with the guidelines stated in the user agreement. Subsequent updates were not reviewed, but New Jersey Online reserved the right to modify any Community Connection site that did not conform to the program guidelines by, for instance, including discriminatory content or commercial goals. If a site was approved, Alford entered that information in the administration tool.[14] Most complete sites were approved.[15]

As a part of her position as community producer, Alford also ran Community Connection's help center. She spent several hours every day answering email and phone calls from actual and potential contributors. Sometimes this involved a quick reply to a person who had forgotten a password, but at other times Alford had to spend a long time on the phone walking a novice Internet user through how to upload an organization's logo—from opening a new browser window to saving the chosen image. Several of the contributors I interviewed mentioned the efficiency of the help center as the key to their having a successful experience with Community Connection.

In an effort to better integrate Community Connection with the rest of New Jersey Online's news content, in early 1999 Alford began systematically to look for links between news stories and related Community Connection sites. That is, every morning she attended the daily meeting of New Jersey Online's editorial department to learn what were the major news stories under consideration. Then, on the basis of that information, she searched the database of Community Connection sites, pulled out sites relevant to various stories, and on each story's page added links to those sites. For instance, if there was a story about a case of child abuse, Alford would place links on that story's page to Community Connection sites offering help to victims and their friends and families. Around the same time, Community Connection launched a forum and a biweekly email newsletter. Although the forum was not monitored, Alford periodically intervened to keep the flow of exchanges going. The newsletter was delivered to all contributors and contained such items as news and upcoming events, sites of the week, Q&As, and profiles of figures in the New Jersey nonprofit sector.

All this work in Community Connection's backstage was complemented by an intense and multifaceted activity aimed at enrolling new contributors. During my fieldwork there was a firm belief at New Jersey Online that this activity was crucial to the continued success of

Community Connection. Although Carla Alford took part in these efforts, the bulk of them were conceived and implemented by Betsy Old, New Jersey Online's director of community relations. Old had vast experience in marketing and in the nonprofit sector, and a comparative lack of experience in technology-related ventures. Hired in September 1998 to fill a new position in the organization, she rapidly developed what she considered to be the general promotional strategy for Community Connection with the premise that "regular, historical marketing and business [strategies] won't work among the nonprofit organizations." "My main strategy for marketing," she told me, "is to come from the bottom and the top at the same time." (interview, March 4, 1999) In other words: "You can't just have a big press conference to make this work, you have to get immersed and . . . get [in the community]." (ibid.)

This dual strategy was manifested in a wide array of practices and initiatives. For instance, during the fall of 1998, New Jersey Online undertook a promotional campaign of Community Connection, which included ads in print, broadcast, and online media, public relations efforts, and direct mail campaigns. These latter involved first developing and then sending copies of a Community Connection brochure (figure 6.5) to more than 13,000 nonprofit organizations. In addition, Betsy Old, Carla Alford, and New Jersey Online personnel took part in a wide array of what they called outreach activities, which involved introducing Community Connection to potential contributors by visiting many nonprofit organizations. Old also did a lot of what she called cyber-PR, which basically consisted of spending numerous hours online, searching for relevant sites, and establishing connections with representatives of those sites that could benefit Community Connection.

Newsroom practices related to Community Connection differ substantively from those commonly found in traditional media, where tasks center on mediating between multiple events, versions, and sources, and the audience. Perhaps nothing captures the essence of this mediation process better than David White's notion of editorial work as gatekeeping[16]: "A story is transmitted from one 'gate keeper' after another in the chain of communications. From reporter to rewrite man . . . the process of choosing and discarding is continuously taking place." (White 1949, p. 384) In contrast, together with those of a gatekeeping character—such as site approval—there was at Community Connection a significant presence of practices that expanded what is normally found in editorial departments, including news contextualization, enrollment of new contributors, database management, technical support, newsletter writing,

It's so easy!

- Connect to: **www.nj.com/cc**
- Fill out the form.
- Follow the simple directions and build a site!

This Program Really Works!

New Jersey Online wants all groups in the state to participate. If you are a:

- School/College/University
- Volunteer/Non-profit Group
- Town/Government Agency
- Civic Group
- Recreational Group
- Religious Organization
- Community Group

You are eligible for a free Web site in the Community Connection program.

In addition to giving your group its own Web site, Community Connection puts you in a statewide, searchable database of New Jersey groups and organizations. The Community Connection makes it easy for others to find your site and learn all about your organization.

Participating groups include:

- Neighborhood Association Network
  http://community.nj.com/cc/neighbors_som
- New Jersey Self-help Clearinghouse
  http://community.nj.com/cc/mash
- Fair Lawn Soccer Program
  http://community.nj.com/soccer/cutters
- New Milford Senior Activities Center
  http://community.nj.com/cc/newmilfordseniors
- Referee Enhancement Focus
  http://community.nj.com/cc/soccerelitereferees
- Brick Terror
  http://community.nj.com/soccer/terror
- Princeton Area Community Foundation
  http://community.nj.com/cc/pacf

IT'S FREE, EASY-TO-USE, AND YOU GET YOUR OWN WEB ADDRESS!
www.nj.com/cc

Benefits

- **It's free!**
- **It's easy.** Groups can create and update their Web sites, without knowledge of computer programming.
- **It spreads the word.** Tell all of New Jersey who you are and what your group does, your group's news and upcoming events, volunteer information and links to your favorite sites.
- **It saves you money.** Post your newsletters online and save on mailing costs.
- **It reaches the right audience.** Recruit volunteers, publicize your group, and promote your services online.
- **It's customizable.** Upload your own logo or graphic image or use our free image library.
- **It lists you in the state's largest online directory.** Sites are automatically entered into a statewide searchable database.

Sign Up. It's Free & Easy!

If your group is not currently a member of New Jersey Online's Community Connection, sign up now by visiting http://www.nj.com/cc or contact us for more information.

For more information, contact:
Community Connection
New Jersey Online
30 Journal Square
Jersey City, NJ 07306
(201) 459-2800
E-mail us at:
community@nj.com

NJO
NEW JERSEY ONLINE

new jersey online
Community
connection
www.nj.com/cc

Figure 6.5
Part of a Community Connection brochure. © New Jersey Online. Reprinted with permission.

and forum hosting. Despite their differences, these tasks converged on a general goal: to facilitate content creation and sharing by a large and heterogeneous set of contributors. If gatekeeping is the image when mediation defines work in an editorial unit, gateopening is a possible alternative with practices focused on facilitation. Workers thus open up the online paper to contributors, turning it into a space for knowledge creation and circulation. This extends the analysis about the material dimension of online newsroom practices presented in chapters 4 and 5 by adding another set of transformations potentially related to the use of technology in online news. The shift from traditional gatekeeping to newsroom routines centered on the facilitation and circulation of knowledge produced by a vast and heterogeneous network of users-turned-producers.

The wide range of practices involved in facilitating the creation and sharing of information by users challenges the boundaries that customarily separate editorial, technical, and commercial domains in print newspapers. One manifestation of this process was in the important technical support function of Carla Alford, which resonates with an even stronger role of the kind played by HoustonChronicle.com's David Galloway in the "At Sea" voyage. The importance of this technical work in Alford's spectrum of newsroom practices and the value it had to keep Community Connection running smoothly suggest the need to rethink the categories often used to make sense of the relationships between computers and work. This issue has long been dominated by the "deskilling-reskilling controversy" in the relationship between computers and work.[17] Whereas proponents of the deskilling position have suggested that computerized tools transfer skills from workers to machines, those adopting the reskilling stance have argued that these tools lead to the emergence of new skill sets. Although expressions of both can be seen in Community Connection, neither captures what Alford's technical support tasks indicate about the uniqueness of distributed construction: that in this alternative regime, media workers and firms "out-skill"—that is, they transfer editorial and technical skills to users located outside these firms' formal domain.

Another expression of this boundary-challenging trend was a questioning of the strong separation between editorial and advertising departments that is so pervasive in the culture of the newspaper industry. Central to the maintenance of this separation is the fact that reporters and editors directly control what they publish, filtering out potential commercial influences. But what happens in the case of an initiative in which the information appearing on its sites is not filtered editorially? I asked variations of this question to many of my interviewees, and their responses ranged over the spectrum of possible answers, revealing how poorly Community Connection fits within established categories of traditional media.[18] Some of the interviewees argued that Community Connection was a marketing initiative. For instance, Peter Levitan, who before joining New Jersey Online was an advertising executive at Saatchi & Saatchi, said to me: "I don't view [Community Connection] as editorial, because editorial would mean that we have a voice. So I think that it's a marketing program based on giving people self-publishing tools." (interview, March 11, 1999)

But to others, Community Connection was an editorial project, albeit one that reshaped the very meaning of editorial content. Among them was Sara Glines, who became New Jersey Online's editor after a long career in print journalism. Community Connection, she told me, "has begun to shape my image of what editorial per se is. . . . Editorial now is web sites built by nonprofit groups, the content in our forums, whatever happens in our chat rooms, in addition to the newspaper stories and our polls." (interview, March 4, 1999)

A third alternative was articulated by Jeff Jarvis, also a seasoned journalist: that Community Connection "is editorial/marketing, it's both." He meant that "the audience's content is as valuable—if not more valuable— as [traditional newsroom content]. . . . So in that sense, all of this community content is editorial, and I want [newsroom] people to have the same feeling of pride and ownership of it." (interview, March 15, 1999) On the other hand, "We could also argue that it's not editorial, because we're not doing it; we're providing the opportunity for it to happen. . . . It's not ours, is yours. So, in that sense, is marketing." (ibid.)

The implications of this blurring of the boundaries between previously more distinct technical, editorial, and marketing domains are many and far reaching, and some of these implications will be discussed in the final chapter. For the moment, I turn my attention to the coordination resources and dynamics related to the information architecture and newsroom practices prevalent at Community Connection.

Coordinating Production

The construction of complex new-media sites requires the collaboration of actors coming from different social worlds. Hence, the coordination of such productive processes is critical to the unfolding of the enterprise.

In chapters 4 and 5, I examined coordination efforts that take place at multiple loci, employ a variety of resources, and exhibit a wide range of dynamics. My study of the Times on the Web's Technology section focused on the connections between print and online newsrooms and paid special attention to the emergence of relational spaces able to foster cross-boundary work. My analysis of the HoustonChronicle.com's Virtual Voyager project looked at the linkages among the diverse occupational groups involved in the web operation and the role of material and symbolic resources in these processes. In view of the distributed construction character of Community Connection, in this section I complement the analysis presented in the two previous chapters by concentrating on a third coordination locus—the intersection of a media organization and its users-turned-producers—and examining the resources and dynamics at play in it.

The unit of production in Community Connection was not New Jersey Online but an ensemble composed by New Jersey Online and a multitude of nonprofit organizations of various sizes and kinds. This heterogeneous ensemble brought a multiplicity of motivations, goals, and evaluative principles to the production process that differed substantially from the more uniform situation of traditional print newspapers. On the one hand, New Jersey Online, a for-profit company, undertook Community Connection with a mix of direct and indirect profit strategies. First, it adopted an advertising strategy for Community Connection different from that prevalent in the rest of the site. The vast majority of New Jersey Online pages contained either paid advertisements or promotions of other parts of the site or both. However, with the exception of Community Connection homepage and its four sections, all of which featured content generated by New Jersey Online, contributors' sites tended to have only messages of the second type.[19] The rationale for this decision was to avoid potential conflicts between nonprofit organizations and particular advertisers. However, New Jersey Online did not relinquish its aspirations to collect revenue directly from Community Connection, but pursued this goal through a "sponsorship model." Instead of having a portfolio of advertisers whose messages would appear on Community Connection pages on a rotational basis, as in other parts of the site, New Jersey Online tried to find one single sponsor for the whole program. Potential candidates for this role were some large companies with a strong presence in the state, whose line of business generated either no or few objections in the nonprofit sector— such as utilities, health care, and financial companies.[20] Second, these "direct" revenue strategies were complemented by an array of indirect ones that contributed to the overall financial performance of New Jersey Online by increasing the site's traffic, raising its visibility in the community, and gaining new users. Sara Glines told me that the business plan behind Community Connection was "to make our site sticky. We want people to come and use it and have a reason to stay and come back tomorrow, and when they are here we want them to feel tied to our site." Hence, "what better way to do that than let them build their site on your site? Now they're going to keep coming back, and will feel certain amount of ownership in New Jersey Online." (interview, March 4, 1999)

The motivations, expectations, and goals of nonprofit organizations were even more varied than New Jersey Online's. My interviews revealed that they wanted the benefit of free web publishing for various reasons, which in turn influenced what they did with their sites and how they evaluated their performance. Some contributors emphasized the importance of Community Connection's local focus. For example, a representative of a small arts group said: "[Community Connection] caters to New Jersey, and our organization is in South Jersey, performs in South Jersey and 100 percent of our [members] live in South Jersey. They also have links to other South Jersey arts organizations." (interview, March 6, 1999) Other groups argued that being part of Community Connection was important in an "information age." For instance, a staff member of an animal shelter said: "Having a web site is the wave of the future, and more and more people are using the Internet looking for information." (interview, February 11, 1999) Some nonprofit organizations that already had a site when Community Connection came along participated to redirect traffic to their primary site, usually building minimal sites and not updating them often. A representative of a cultural group said: "We are not publishing to their [New Jersey Online] web site; it is only a page to direct traffic to our web site." (interview, February 26, 1999) Yet other interviewees said their groups participated for the benefits that could be derived from being part of a statewide directory. For example: "The indication that it is part of a searchable database that New Jerseyans can turn to also made it appealing." (interview, February 22, 1999) Unsurprisingly in view of the usually limited financial resources of many organizations of this sort, several respondents emphasized the free nature of Community Connection as a reason for contributing to it. "In our organization," said one interviewee, "we look at the books at the end of the year, and we have zero in the bank. Web access through Internet Service Providers is expensive, cold, and impersonal, and not user friendly. We have members who can upload and edit the information, but we cannot pay to have someone do this for us. So Community Connection . . . is a good resource for groups like us . . . because it's free!" (interview, March 6, 1999)

In chapters 4 and 5 I showed the value of looking at the coordination of production in new media as articulation work—the practices that "establish, maintain or break" (Gasser 1986, p. 211) the linkages across actors' "primary" tasks and also across the units to which they belong— and analyzed various resources employed to this end as well as the dynamics of the whole process. In this chapter I continue along the same lines by reflecting on the articulation of Community Connection's heterogeneity of goals, means, and worldviews into a "coordinated cacophony of construction" (Galison 1997, p. xxi). Several resources were deployed by New Jersey Online and the participating nonprofit organizations. First, in shifting from gatekeeping to gateopening, newsroom practices focused partly on managing the flow of information from a multiplicity of content providers. Among other important elements that contributed to the success of such an endeavor were the outreach efforts of Betsy Old and, to a lesser extent, Carla Alford. I observed many of these efforts and was always surprised by the extent to which they were as much about promoting Community Connection as about hearing what existing and potential contributors thought of the program and of ways to improve it. Some of these suggestions triggered innovations later on, such as the planned additions of forums and interactive maps to contributors' sites. In addition, because she joined Community Connection after several years in the nonprofit sector, Old was particularly well qualified to understand the concerns of nonprofit organizations and a for-profit such as New Jersey Online, and address each constituency in its own language.

In chapter 4 we saw the use of prototyped and finished voyages as boundary objects to bridge the needs of editorial and design personnel in the making of Virtual Voyager. A second key resource in the coordination of information production at Community Connection was not one, but a collection of objects: the publishing and administration tools, several sections of the Community Connection sites authored by New Jersey Online personnel (the instructions, the FAQ, the user agreement, and so on), the newsletter, the forum, and, last but not least, the groups' sites themselves. In a sense, each could be seen as a boundary object. For instance, the publishing tool was robust enough to maintain a common identity across sites, yet flexible enough to satisfy the information needs of an array of nonprofit organizations publishing a variety of content, as well as of New Jersey Online's need to house all the content in a single section of the site. This was certainly not a random occurrence. The tool was consciously built to achieve these multiple goals and was subsequently refined through the attention that New Jersey Online personnel paid to user feedback. However, much like appreciating the performance and music of an orchestra involves looking beyond each instrument individually, understanding the role played by this collectivity of artifacts requires a holistic view. In Bowker and Star's terminology (1999, pp. 313–314), this collective became the "boundary infrastructure" of Community Connection: "Because they deal in regimes and networks of boundary objects—and not of unitary, well-defined objects—boundary infrastructures have sufficient play to allow for local variation together with sufficient consistent structure to allow for the full array of bureaucratic tools—forms, statistics, and so forth—to be applied. Even the most regimented infrastructure is ineluctably also local."

This kind of "localizable" infrastructure was crucial in a distributed construction initiative involving producers who belonged to a multitude of organizations. It enabled the common production activities of parties with diverse backgrounds, resources, and objectives. In their various practices, nonprofit organizations of all sizes and kinds and New Jersey Online customized this boundary infrastructure to their particular needs in such a way that they nonetheless contributed to a common project. Because of this customization, coordination proceeded without homogenizing actors' discourses and practices. Moreover, the information creation and circulation practices of Community Connection were so vast and heterogeneous that no single boundary object would have sufficed to coordinate them. That is why a multifaceted boundary infrastructure was needed for this heightened flexibility, even if at the cost of an equally increased degree of complexity in infrastructure design and management.

This last issue begs a further question: how can we characterize the organizational form of a production system which inscribes users as active co-constructors, centers the editorial function around the facilitation of content exchange, and houses a multiplicity of information flows? A comparison with the dominant organizational form in the print newspaper industry will help us to understand the specifics of Community Connection in particular, and distributed construction in general. Like firms in many other industries, print papers create editorial products by turning inputs from suppliers—information from sources, official records, wire services, and so on—into outputs for consumers—stories for readers. But, unlike many of these firms, dailies generate a new product every 24 hours. Moreover, these products are often about complex and unpredictable events. To cope with these conditions, the editorial function has been constituted as mediation work, the product defined as a unidirectional flow of generalized content, and readers inscribed as content consumers rather than producers. To manage such a production system, dailies, like many firms since the nineteenth century, have become organizational hierarchies with centralized authority and relations of dependence among the various levels.[21] Many studies have documented the dominance of this organizational form, even evoking images of mass-production. "Except for extraordinary circumstances . . . news is processed in a way that satisfies economic and technological constraints comparable in their rigidity to those of other mass-production processes." (Roshco 1975, p. 111)

Community Connection presented a different organizational form derived from the fact that many users not only consumed the output but were also in position to supply the input, and that these "co-producers" did not belong to New Jersey Online but to a vast and heterogeneous array of organizations. Thus, productive activities in Community Connection were organized around a "polyphonic regime of worth" (Stark 1996). New Jersey Online pursued Community Connection for reasons as diverse as free content, traffic growth, new users, higher visibility in the community, and financial revenue. Nonprofit organizations contributed to this program for an even more varied array of motives, ranging from reaffirming their local identity and community presence, to being part of the "information age," to simply redirecting traffic to an already existing site. The co-existence of these various organizing rationalities in a single unit of economic action gave Community Connection a behavioral versatility that distinguished it from traditional bureaucracies, turning it into what Teubner (1993, p. 57) has called "many-headed hydras" or entities with "polycentric autonomization . . . [able to] act collectively, not through a single action center, as is typical for the classical corporation, but through a multiplicity of nodes."

In view of the heterogeneity of organizing rationalities and institutional affiliations, the relationships among the members of this ensemble were not of dependence and control but of interdependence and trust. For instance, New Jersey Online depended on nonprofit organizations to provide appropriate content and promote the service among their constituencies, whereas nonprofit organizations depended on New Jersey Online to administer the service adequately and enroll as many new contributors and users as possible. Rather than a hierarchy, such an organizational form can be better characterized as what Stark (2001, p. 75) has called a "heterarchy," since "whereas hierarchies involve relations of dependence and markets involve relations of independence, heterarchies involve relations of interdependence."[22] In these new organizational forms, authority does not reside at the top and decrease as it goes down the hierarchy, but is distributed more evenly—though not entirely—throughout the collective. Thus, even though New Jersey Online had the right to eliminate a site if it did not conform to the stated guidelines, nonprofit organizations could also move their sites to other competing services at a relatively low cost and without experiencing any retaliation. A similar organizational form is also prevalent in forums and chat rooms where neither the contributors nor the online paper can exert unilateral control over content production, and all parties depend on each other to keep the process going.

The emergence of these alternative organizational forms relates to the boundaries of economic action. Although heterarchies are, to use Sabel's image (1991, p. 25), "Moebius-strip organizations" ("It is impossible to distinguish their insides from their outsides"), most research on new organizational forms has implicitly kept the separation between production and consumption by focusing on economic action happening within and among firms. However, the production locus in distributed construction resides at the intersection of firms and users, thus blurring not only the boundaries among organizations but also those between organizations and their publics. To make matters more complex, Community Connection has been part of an online paper in which a more hierarchical form has been dominant. Thus, when distributed construction initiatives happen within traditional media operations, they result in the co-existence of diverse organizational forms.

Concluding Remarks

Community Connection expresses a new regime of information production in online newspapers I call "distributed construction." This regime results from combining an artifact configuration inscribing users as content co-producers and enabling a multiplicity of information flows, with newsroom practices mixing facilitation and mediation tasks and a heterarchical organizational form. These transformations involved in distributed construction are interdependent, meaning that alterations in one domain are tied to changes in the others. For instance, inscribing users as co-producers was deeply intertwined with Carla Alford's and Betsy Old's gate-opening tasks as well as with the emergence of a heterarchy to manage interdependence and multiple rationalities. These practices represent a marked departure from the typical case of content production in print and most online newspapers. (See table 6.1.)

The notion of distributed construction is inspired by research on "distributed cognition."[23] In the same vein that distributed cognition has argued that the process of knowing does not take place inside an individual's brain but emerges from the interactions with the social and material environment, the thrust of distributed construction is that, given certain conditions, content production in new media does not happen inside a firm's newsroom but results from the interactions with users. In challenging the separation between production and consumption, distributed construction also resonates with production systems that were common in the early nineteenth century, such as "putting-out."[24] However, unlike these prior systems that regulated relationships between factory owners and workers that were alternatives to the traditional employment contract, distributed construction illuminates new engagements between media organizations and consumers who contribute to the production process while making a living in some other way.

Distributed construction challenges dualities that have been pervasive in understanding the relationships between communication, technology, and organizations in the media industry: construction and effects of messages, artifact development and use, and content production and consumption. When a nonprofit created a Community Connection site, it

Table 6.1
Comparing the dynamics of centralized and distributed construction.

	Centralized construction	Distributed construction
Artifact configurations	User inscribed as consumer	User inscribed as producer
Work practices	Centered on mediation	Mostly focused on facilitation with some mediation
Information flows	Mostly unidirectional	Multiple directions
Organizational forms	Hierarchy	Heterarchy

was at the same time source and destination, builder and user, and pro-ducer and consumer; and the same was true with New Jersey Online, always oscillating from source-builder-producer into destination-user-consumer. This does not mean that it is no longer possible to distinguish between the members of each duality, but rather that they get "de-reified" (Berger and Luckmann 1966) as they turn from givens into situated achievements.

Beyond the specifics of Community Connection, the analysis pre-sented in this chapter extends developments concerning two analytical insights elaborated in chapters 4 and 5, and adds a third one about the social study of new media. First, the construction of Community Connection reinforces the idea that technical practice and, more gener-ally, technical considerations are part and parcel of work routines in online newsrooms. The notion that materiality matters was manifested in multiple actions undertaken by newsroom actors, from Alford's technical support activities to the design of the publishing tool and the nonprofit organizations' sites, to the strategy of promoting Community Connection as technically unsophisticated. Furthermore, this examination of Community Connection expands our understanding of the spectrum of possible transformations in traditional newsroom patterns that may take place in relation to appropriating the interactive capabilities of online technologies: the shift from typical gatekeeping practices to those oriented toward the opening of the newspapers' gates by facilitating information creation and exchange by an ensemble of users-turned-producers.

This first insight leads to the second one: the idea that the character of new media arises from the locally contingent production practices that appropriate technical capabilities, and that these dynamics are made invisible by a focus on the supposedly unique attributes of new-media artifacts. In chapter 5, I argued this point in reference to the practices that brought forth immediacy and hypermediacy in Virtual Voyager, and the parallel absence of such practices in the Technology section. The making of Community Connection allows us to underscore this general analytical point with regard to interactivity. Understood as the possibility of extending traditional mass media's one-to-many information flow to a many-to-many one, interactivity has been another technical attribute that has often been used to differentiate new from old media. In the case of online newspapers, this focus on interactivity as a product has surfaced in research that addresses such issues as to what extent various sites are or are not interactive, what users do with the available interactive features, and how these features may affect broader societal dynamics.[25] Concerns about interactivity as product have also permeated general reflections about the new-media landscape, such as the following (Poster 2001, p. 48): "The Internet extends the figure of the producing consumer of earlier technologies and makes this the very principle, the automatic operation, of every communication."

The problem with this focus on product attributes is that it misses variations in interactivity that depend to a significant extent on the practices that produce these attributes. In other words, differences in degree and kind of interactive communications abound in online newspapers, and the web more generally, and result partly from local practices and not from the automatic operation of technical features. For instance, making sense of the differences in the realization of the web's interactive capabilities between New Jersey Online and the Times on the Web and HoustonChronicle.com has to take into account local variations in production dynamics, such as New Jersey Online's populist vision of the online environment and enactment of newsroom practices that deviate strongly from traditional gatekeeping routines. Thus, although online environments are potentially more interactive than print, radio or television, the differences and similarities among the actual media products can be understood better in relation to their processes of production.

This book focuses on the production of online newspapers, but an examination of the dynamics involved in the heightened role of users in Community Connection, together with some observations about user practices related to the Technology section and Virtual Voyager, yields the third analytical insight, this one into the consumption of new media. During the past two decades, scholarship in computer-mediated communication has generated valuable knowledge about online behavior.[26] For the most part, knowledge in this area has been generated by research that splits online behavior from its offline context and treats it as a stand-alone object of inquiry. Although it has been useful in reducing the complexity of relatively new and unknown phenomena, this mode of inquiry has been, nonetheless, limited because it introduces an artificial division into what are two intertwined domains. According to Haythornthwaite (2001, p. 363), a focus on "[computer-mediated communication] versus face-to-face, online versus offline, and virtual versus real" has "perpetuated a dichotomized view of human behavior . . . [though] a growing body of research is now examining more integrative views of [computer-mediated communication]."

This alternative body of research has mostly looked at the interpenetration of online technology use and various facets of everyday life.[27] My account of Community Connection suggests that we can also gain by examining another important area of integration: the predominantly offline construction of the artifacts that enable online communication and how this shapes, and is shaped by, their users' actual practice. For instance, in comparison with a hypothetical research design that concentrates exclusively on nonprofit organizations' online communications, a more textured understanding of how and why nonprofit organizations adopted Community Connection as opposed to other free web-publishing options emerged from considering New Jersey Online's choices about the construction of a technical infrastructure that targeted technically limited users. Along similar lines, the kind of online involvement of users in Virtual Voyager's "At Sea" was dependent on such offline technical and communication factors as modifying the site's structure to make visible the exchanges between sailors and users and having Galloway spend a significant portion of his time allocated to this project in managing the flow of email messages addressed to the sailors. Therefore, linking people's behaviors in online environments to the social and technical factors that shape the mostly offline construction and management of these environments should lead to a more encompassing picture of the dynamics of new media.

7

"When We Were Print People"

In this book I have sought to understand the practices whereby actors situated in established media appropriate novel technical capabilities, and the new media that result from these practices. To this end, I have examined how American daily newspapers have dealt with the promises and perils of consumer-oriented electronic publishing, with special attention to the construction of online newspapers on the web. Theoretically, I have drawn from technology, communication, and organization scholarship to illuminate the material, editorial, and work dimensions of these practices, as well as the intricate relationships that tie these dimensions. Methodologically, I have embedded a contemporary focus within a historical perspective to situate descriptively thick but temporally and contextually circumscribed case studies within more extended and broader dynamics.

In the first section of this concluding chapter, I will recapitulate the main empirical findings and analytical insights introduced in the previous chapters. The empirical findings are encapsulated in the elucidation of two patterns of innovation that have influenced newspapers' appropriation of nonprint alternatives. First, print papers have enacted a culture of innovation that led them to react to social and technical developments rather than more proactively contribute to these developments, focus on protecting the print franchise rather than on prioritizing nonprint publishing, and emphasize smaller but more certain shorter-term gains rather than potentially larger, but less certain, longer-term benefits. Second, a comparative analysis of the case studies presented in chapters 4–6 reveals that the paths pursued by these initiatives that attempted to take advantage of the web's distinctive capabilities have been shaped by three factors: the relationships between the print and online newsrooms as either close or distant, the representation of the intended user as either consumer or producer of information and either technically savvy or unsavvy, and the character of online newsroom practices as either reproducing editorial gatekeeping or generating alternatives to it. Different combinations of these factors have led to different innovation paths in online newsrooms.

Making sense of these patterns of innovation furnishes analytical insights about the construction, products, and use of new media. First, in contrast with the silence about the role of technology in the majority of social studies of newsmaking, my research shows that technical practices and considerations are central in the editorial work that goes into the construction of information in online newsrooms. Then, as opposed to the neglect of production dynamics in most scholarship on the differences between old-media and new-media products, my study indicates that these dynamics are critical to understanding the alternative forms that new-media products may acquire. Furthermore, contrary to the split between online and offline domains in most computer-mediated communication research, my analysis suggests that it is important to account for the largely offline shaping of the content and artifacts that enable users' online experience.

Then, in the second and third sections of this chapter, I will draw from these findings and insights to offer reflections about two critical trends in the recent shaping of the new-media landscape, and whose influence is likely to grow in the near future: the dynamics of media convergence, and the reconstruction of news in the online environment. Regarding media convergence, I will show that the dominant discourse about it has suffered from looking at convergence as an end state, concentrating on its supposedly revolutionary attributes and neglecting the role of pre-convergence differences affecting post-convergence trajectories. In contrast, I will argue that the notion of emerging media, with its associated emphasis on history, locality, and process, provides an important corrective to this discourse by making visible the evolutionary and situated dynamics that mold the diverging paths along which new media unfold. Concerning the character of online news, I will suggest that a larger number of groups seem to have a higher degree of agency in shaping the news than the typical case of print and broadcast media, and that this puts a premium on coordination across the boundaries that separate these different groups. This, in turn, appears to influence the creation of news that are more user-centered, communicated as part of ongoing conversations, and with a micro-local focus.

Innovation in Online Newspapers

In chapters 2 and 3, I showed that American dailies' pursuit of alternatives to consumer-oriented print publishing shifted from exploring an array of possibilities in the 1980s, to settling on the web circa 1995, to hedging with their web sites during the second half of the 1990s. First, newspapers tinkered with an array of technical and content options, from videotex to audiotex systems, and from editorial to transactional material, and closely examined the commercial feasibility of these options. After a decade of exploration, and although they never completely ceased to experiment with other information environments, newspapers concentrated the vast majority of their nonprint efforts on the web. Once they began settling on the web, newspapers hedged by proceeding in many directions, from merely reproducing their print content, to enhancing it with the addition of new content or technical capabilities, to developing a largely novel suite of information products.

This trajectory from exploring to settling to hedging expresses a culture of innovation marked by reactive, defensive, and pragmatic traits. Newspapers usually followed players that moved first in relation to potentially relevant technical and social developments. For example, the move from online services to the web in the mid 1990s partly resulted from perceived changes in consumer preferences and took place after early entrants had acquired a leading market position, even though newspapers had been dealing with online communications since before these new companies were even formed. Then, when they acted, newspapers undertook projects with as close an eye on what they meant for their print franchise as on establishing a presence in a new territory. Furthermore, in these projects, newspapers were often more concerned with the short-term success of products that related to what decision makers saw as their "core business," than with the uncertain possibilities of more experimental artifacts that could only pan out in a longer time horizon. Thus, for example, videotex papers started because decision makers saw them as a possible threat to the print franchise, and were terminated because that threat did not appear to be significant in the near future, not because they did not work technically or hold promise as an alternative communication space.

This culture of innovation has had mixed consequences. On the one hand, newspapers' nonprint endeavors have generally proceeded in a slower and more conservative fashion than those of organizations less tied to traditional media. This difference becomes evident when one

compares leading online papers such the Times on the Web, washing-
tonpost.com, and ChicagoTribune.com with born-on-the-web sites such
as CNET, Feed, Monster.com, and Geocities, which have competed with
online papers in the markets for news, features, classified ads, and user-
authored content, respectively. On the other hand, the cumulative transformations after two decades have been remarkable, as can be seen, for
instance, in the difference between the products and services offered by
these leading online papers and their print counterparts. Therefore, I
have argued that in their pursuit of permanence, undertaking innovation
to stay the same, newspapers have nonetheless ended up generating
substantial change.

Within this broad culture of innovation in print newspapers, the paths
followed by the New York Times on the Web's Technology section,
HoustonChronicle.com's Virtual Voyager, and New Jersey Online's
Community Connection, three initiatives that have tried to take advantage of the distinctive capabilities of the web, have been shaped by three
factors: the relationship between the online and print newsrooms, the
inscription of a vision of the intended user in the technical and communication content of the product, and the character of newsroom practices as either reproducing editorial gatekeeping or enacting alternatives
to it. Different combinations of these factors have led to different online
innovation paths and different media artifacts.

The first factor centers on the relationship between print and online
newsrooms. Print newsrooms have been around for a long time, which, if
nothing else, carries substantive moral authority when it comes to dealing with the uncertain prospects of constructing media artifacts in an
unknown information environment. In addition, modern print newsrooms have developed highly standardized procedures, furnishing "tried
and true" templates to initially approach the editorial work related to the
web. Furthermore, during their first 5 years of existence, most online
papers on the web were largely being funded by financial resources generated in the print business, which also gives significant symbolic weight
to the print newsroom. This asymmetry between print and online newsrooms influenced innovative efforts in the three case studies in the following way. On the one hand, the more extensive the efforts undertaken
to align print and online newsrooms, the more reproduction of print's
ways of doing things in the online environment. The Times on the Web's
Technology section is an example of this option: its editors spent significant portions of their routines neither on story assignment or copy editing, but coordinating with their counterparts at the relevant desks in the

print newsroom.[1] On the other hand, the less extensive such work of alignment, the less "repurposing" of print's world in the nascent online domain. Both the *Houston Chronicle's* Virtual Voyager and New Jersey Online's Community Connection are illustrations of this alternative: neither Virtual Voyager nor Community Connection staff had much regular contact with the newsrooms of their affiliated print papers, and no major effort was made to link online and print products more generally.

The second factor relates to the inscription of a representation of the intended user in the media artifacts produced by online newspapers. That is, the people who participate in the construction of information in online newsrooms have an idea of what kind of users they would like to reach and inscribe this idea in technical and communication domains such as interface design, media choice, and information flows. Two dimensions of this inscription appear as particularly relevant in the present cases: users' technical expertise and their position as either consumers or producers of content. One way in which issues of technical expertise played out in the cases is that building technically unsophisticated products was seen as crucial to continue with print papers' broad and general audience; conversely, a high degree of technical sophistication meant targeting primarily the "lead user/early adopter" public. On the one hand, the more producers inscribed users as technically unsavvy, the more they communicated via text and still images. This situation is illustrated by the dominance of simple interfaces and textual information in both the Technology section and Community Connection. On the other hand, the more producers inscribed users as technically savvy, the more they took advantage of the web's multimedia capabilities, with Virtual Voyager being a prime example of this alternative.

The other relevant dimension of user inscription revolves around issues of information flows. Newspapers have long represented their readers as consumers of content and constructed an artifact that reflects this image by leaving little space for readers to voice their opinions alongside those of reporters and editors. This preference has also been influential in the direction of innovation in their online counterparts. On the one hand, the more users were conceived as wanting to produce, not just consume, the content appearing on the online paper, the more enactment of multiple information flows. The most evident illustration of this was Community Connection, in which users contributed most of the information available on the site. On the other hand, the more users were seen primarily as consumers of content, the more reproduction of print's "we publish, you read" mode. Examples of this possibility were the

Technology section's compartmentalization of forums from articles and columns, and, with the exception of "At Sea," the relative absence of user involvement in Virtual Voyager.

Issues of information flows relate to the third factor that shaped newsrooms' online innovation paths, which has to do with the character of the editorial function. All occupations and professions have certain traits that distinguish them and make them stand apart as a recognizable domain of activity. In the case of modern journalism, one such key attribute is the notion of gatekeeping. The idea that editorial work is about mediating between events and consumers is transmitted anywhere from journalism school to on-the-job socialization and has influenced print's disregard for reader-authored content. Hence, it is not surprising to find that it has played an important role in print's appropriation of online's multi-directional capabilities from the 1980s' videotex to the 1990s' web initiatives. We saw variations of this theme in the three case studies. On the one hand, the more the editorial function was configured around gate-keeping tasks, the more reproduction of print's one-to-many message flows. The Technology section's enactment of traditional journalism routines is an illustration of this option. In contrast, the more the editorial function was configured around alternatives to gatekeeping, the more a multiplicity of information flows was enacted. Community Connection's adoption of user-authored content was coupled with newsroom practices centered on facilitating and managing multiple streams of information flows. In a sense, the factors concerning the character of newsroom practices and the inscription of users' consumption and production position were the two sides of the interactivity coin, one focusing on the work process and the other on the beneficiary of its products.

The combination of the three factors evokes an image of patterned diversity. (See table 7.1.) This image has at least two implications for understanding the dynamics of online papers. First, there has not been one but diverse innovation paths in the online newsrooms studied. Second, the direction and pace of these paths has not been random, but patterned in relation to factors that mark both continuity with the world of print newspapers and also the possibility of substantial deviations from it.[2] This image of patterned diversity suggests that there has not been just one embodiment of online journalism, nor have the various alternatives progressively converged toward one form (an issue with far-reaching implications that will be addressed more fully in the next section).

Making sense of these patterns has yielded three general analytical insights about the construction, products, and use of new media.

Table 7.1
Factors that shaped the online paths pursued in the three case studies: the New York Times on the Web's Technology section, HoustonChronicle.com's Virtual Voyager, and New Jersey Online's Community Connection.

	Relationship between print and online newsrooms	User inscriptions	Character of newsroom practices
Technology section	Extensive articulation of alignment	Technically unsavvy consumer	Traditional print reporting and editing
Virtual Voyager	Limited articulation of alignment	Technically savvy consumer	Multimedia reporting and editing
Community Connection	Limited articulation of alignment	Technically unsavvy producer	Content management and facilitation

First, in contrast with most depictions of newsroom dynamics, my research has shown that materiality matters in online newsrooms. That is, online newsrooms appear as sociomaterial spaces in which technical considerations affect who gets to tell the story, what kind of stories are told, how they are told, and to what public they are addressed. For example, people in charge of the Technology section were aware that to reach as wide an audience as possible, they had to tell stories via technically unsophisticated media artifacts, which also affected staffing decisions, such as the hiring of freelancers with limited multimedia background, and their subsequent use of rather traditional journalistic formats. The findings also reject deterministic explanations of online technologies transforming journalism in fixed directions, highlighting instead the role of contextually dependent processes and outcomes. That is, decision makers in the innovations I studied shared at least a basic awareness that there was a range of options in interactivity and multimedia, but took advantage of them differently as a result of local dynamics. Some specific changes from traditional journalism that were associated with the appropriation of online technologies include the interpenetration of print, audiovisual, and information systems practices in the making of multimedia products, the de-reification of media options that occurs when actors can choose whether to use text, audio, video, and animation to tell a story, the challenges to established occupational identities that happens when print journalists appropriate the alternative communication possibilities available in online environments, and the rise of an editorial function geared to the facilitation and management of user-authored content.

The second analytical insight suggests that attempting to understand what is and is not distinctive about the new media through an exclusive or predominant focus on their products, with no or little attention to their processes of production, runs the risk of attributing either cultural or technological necessity to locally contingent outcomes. More concretely, my analysis has shown that neither remediation is a cultural nor interactivity a technological necessity of new media, but that both were enacted to varying degrees in the three case studies as a result of different combinations of locally contingent factors such as the above mentioned relationships between print and online newsrooms, inscriptions of a vision of the intended user in the media artifacts, and characterizations of newsroom practices.

The issue of interactivity also leads to the third analytical insight, which centers on the role that the offline construction of new-media artifacts have in users' online experience. In contrast with the limited understanding of online behavior that arise from overlooking its ties to relevant offline processes, this study indicates that a more comprehensive and textured picture of the use of new-media products results from taking into account the offline dynamics that influence this use. For example, the enactment of a dialogue between sailors and audience in "At Sea," and the appropriation of online publishing by nonprofit organizations in Community Connection were dependent on offline dynamics, such as artifact designs, user inscriptions, and newsroom routines.

Building on these empirical findings and analytical insights I have argued that innovation in online newspapers has unfolded by weaving the sociomaterial infrastructure of print with the novel possibilities associated with developments in information technology. Newspapers have neither stood still in the midst of major technological changes, nor incorporated them from a blank slate, but appropriated novel capabilities such as multimedia, interactivity, variable publication cycles, and simultaneously micro-local and global reach from the starting point of print's culture. This appropriation has not been uniform either, with significant variation resulting from combinations of different initial resources and goals, contextual factors, and process dynamics. The notion of emerging media is an attempt to capture the idea that new media emerge from merging existing infrastructures with novel technical capabilities in an ongoing process shaped by initial conditions and local contingencies.

These findings and insights also highlight the value of history, locality, and process in the study of new media. Historicizing what are usually symbols of the future helps us to understand the extent to which the past influences the present and to evaluate the sources and implications of discontinuous trends. Localizing media practices permits us to make sense of the factors that shape how actors situated in different contexts but experiencing relatively similar environmental pressures and opportunities often appropriate the same technological developments in varying ways and with varying results. Attending to process dynamics contributes to making visible the recurrent calibration between original goals and actual enactment of media practices and accounts for the often unforeseen consequences of such practices.

From Media Convergence to Emerging Media

The proliferation of options concerning the communication, technological, and organizational dimensions of online news invites reflections about the issue of "media convergence." This has been one of the most pervasive but least empirically examined notions in discourse about new media since the 1980s.[3] As Jenkins (2001, p. 93) has argued, "few contemporary terms generate more buzz—and less honey." Briefly put, the notion of convergence has been usually employed to refer to the delivery of content and services previously provided by several media to a single artifact, often a networked computer. "The prophecy of convergence is this: television sets, telephones, and computers—and the networks that bind them—are or will become the same. The Internet will be all." (Owen 1999, p. 16)

There have been three interconnected themes that cut across most treatments of convergence. The first is a focus on the products of convergence, with much less attention spent on understanding the processes that create these products. That is, scholars have focused on convergence as an end state, and comparatively overlooked the processes whereby this end state can be reached. According to Negroponte (1996, p. 18), "when all media is digital . . . bits commingle effortlessly." The second is an emphasis on what is new and unique about these products, with a parallel disregard for the ways in which they fuse old and new and common and unique traits. In connection with these themes, researchers have tried to understand media convergence by either characterizing the logic behind the new products or assessing their editorial, historical, organizational, and regulatory consequences.[4] The third theme derives from the first two: perhaps the most discussed unique trait of convergent products is that, as a result of the combination of previously distinct analog technologies into a single digital domain, there is an erasure of the differences among the originating media forms as they blend in the new-media space. This theme has been expressed with a telling image by Greenstein and Khanna (1997, p. 201) in their discussion of the implications of convergence for corporate strategy: "The Allegheny River forms in Pennsylvania, loops into western New York, and then flows southward into Pennsylvania. The Monongahela River forms in West Virginia and flows northward into Pennsylvania. At the very center of Pittsburgh, these two major waterways converge and become the Ohio River. Until they converge, geographers and everyone who uses them can distinguish them; where they converge, there is neither one nor the other but only a new thing: the Ohio River. Industries that have been distinct historically, even as recently as a decade ago, converge in an analogous way."

Drawing from the empirical research featured in chapters 2–6 and building on the insights of a handful of dissenting views,[5] I would like to suggest that the notion of emerging media helps to make visible what is left unexplored by the dominant discourse around convergence.

First, looking at the process of producing convergent media shows that, inverting Negroponte's assertion, bits commingle effortfully. That is, making the bits of a multimedia artifact such as Virtual Voyager's "Asleep at the Wheel" commingle involved an inter-unit, inter-occupational effort, including such disparate tasks as searching and compiling documentary sources, designing an interface adequate for both navigation and storytelling purposes, and dealing with the logistical complexities of gathering, editing, and producing multimedia content on the road. Similarly, to construct the hybrid of newspaper, library, database, and public forum represented by the Technology section's "Microsoft as Monopoly" meant combining multiple reporting efforts by print and online newsrooms with multifaceted production work by online staffers. Moreover, the lack of connection between the articles authored by editorial personnel and the content provided by users in the "Microsoft as Monopoly" dedicated forum, and the loss of the potential gains to both journalists and users from a more intense exchange of views, shows that, in the absence of efforts to make them commingle, bits do not commingle at all.

Second, looking at the ways in which convergent products emerge from the merging of old sociomaterial infrastructures with new technological capabilities sheds light on the evolutionary dynamics of both processes and outcomes. For example, videotex, which was seen as highly dissimilar from traditional media by actors in the early 1980s, was appropriated by print newspapers in a way that reproduced many facets of print culture. In Viewtron, the most ambitious videotex enterprise by an

American paper, for instance, decision makers ignored multiple signs of interest in user-authored content because this content did not fit with the traditional newspaper world. This is not to say that videotex newspapers were the same as their print counterparts, but they were not wholly unlike them either. More than a decade later, similar dynamics characterized even the more cutting-edge initiatives by online papers. For example, Virtual Voyager and Community Connection enacted print-based practices such as unidirectional information flows in the former and text-based communication in the latter. All this points to the critical need to take into account the merging of old and new to map how various convergent media forms unfold in the new-media space.

Third, focusing on the evolutionary ways in which actors situated within established media organizations appropriate new technical options underscores, unlike the rivers alluded above by Greenstein and Khanna, the heterogeneity of products and production contexts associated with this process and the role of past trajectories in future paths. If, as I have shown, there are significant variations within a single industry, there is reason to speculate that there might be even more variation across the paths pursued by actors situated within, for instance, print, television, and born-on-the-web settings. Differences in factors such as occupational and professional routines and values, technical infrastructures, organizational and industrial traditions, and regulatory regimes are not likely to disappear as a result of the availability of digital networked information environments. On the contrary, these factors have already influenced multiple embodiments of convergence, and it is likely that this will continue in the foreseeable future. Furthermore, although, as I have argued in chapters 2 and 3, the construction of online newspapers has challenged the boundaries that separate the once neatly divided territories of print, broadcast, and telecommunications, this does not mean that this will lead to a world without borders, but probably to one with different boundaries and more cross-boundary work. As Flichy (1995, p. 173) maintains in his historical survey of communication technology, "technological integration will result far more in the movement of borders between the various media than in the removal of these borders."

As a result of making visible these often underexplored dimensions, the notion of emerging media invites one to ask questions about issues of process, and to recast the inquiry of the products and consequences of convergence. Investigating the process dimension means addressing such issues as how the legacy of past endeavors influence present initiatives, what local factors are relevant in shaping the various realization of networked computing in multiple settings, and the extent to which the dynamics of the processes whereby actors appropriate new technologies—for instance, planned versus improvised change—influence the various products that result from them. Taking a new look at products and consequences entails such issues as focusing on identifying the various ways in which new-media products can diverge, accounting for why this might be the case, and analyzing the various communication, technological, organizational, and legal implications of the diverse paths along which emerging media unfold.

The Reconstruction of News in the Online Environment

The news is a culturally constructed category, as it has been demonstrated by work in two traditions of inquiry. First, social histories of the press have illuminated the institutional and technological factors that have shaped the news over the past 200 years. Schudson (1978) has examined the institutional transformations linked to the emergence of the modern notion of news in the United States and argued that it was invented by the penny papers in the 1830s as a reaction to the growth of a democratic polity, a market economy, and an upwardly mobile middle class. According to Schudson (ibid., pp. 22–23), these papers "began to reflect, not the affairs of an elite in a small trading society, but the activities of an increasingly varied, urban, and middle-class society of trade, transportation, and manufacturing."

Blondheim (1994, p. 38) has emphasized the role of technological changes, and shown that the development of the telegraph informed the evolution of the news in the second half of the nineteenth century: "The telegraph, by increasing the speed of news and making its continuous transmission possible, broke down the reporting of developing news stories into smaller and more frequent segments. . . . [It] also promised to expand the scope of the news. Now it would be news from all over the country, not merely local events, that could capture the attention of the public and create expectations as to future developments and the resolution of events."

The second tradition of inquiry, ethnographic studies of news production, has shed light on the local contingencies that influence the reporting of current events. The fundamental premise of this tradition of inquiry has been that rather than having an essential quality to it, the news, as Gieber (1964, p. 180) has put it, "is what newspapermen make it." Carey (1986, p. 160) has summarized the main contributions from these studies by stating that the news is not "some transparent glimpse at the world. News registers, on the one hand, the organizational constraints under which journalists labor [and] on the other hand, the literary forms and narrative devices journalists regularly use to manage the overwhelming flow of events."

Insofar as what counts as news is influenced by social and technological developments, what can the accounts presented in chapters 2–6 suggest about the potential reconstruction of news in the online environment? In the remainder of this section I will adopt the nonessentialist premise that the news in the online environment is what those contributing to its production make it. Although some of the groups seen as having agency, such as forum participants in the Technology section, and the content deemed as relevant, such as nonprofit organizations' publications in Community Connection, would not be included in traditional definitions of news makers and news products, I would like to underscore the value of "following the actors" (Latour 1987) to reflect about the possible reconstruction of news online. In so doing, at least two transformations appear to distinguish the production of new-media news from the typical case of print and broadcast media: the news seems to be shaped by a greater and more varied groups of actors, and this places a premium on the practices that coordinate productive activities across these groups. This, in turn, seems to influence the content and form of online news in three ways. The news moves from being mostly journalist-centered, communicated as a monologue, and primarily local, to also being increasingly audience-centered, part of multiple conversations, and micro-local.

In the online environment, a greater variety of groups of actors appear to be involved in, and have a more direct impact on, the production process than what is typically accounted for in studies of print and broadcast newsrooms. These studies have tended to focus on the work of editors and reporters. Based on the analysis presented in the previous chapters, it is reasonable to speculate that at least four additional groups of players may be having a growing degree of agency in new-media news production. First, in the case of the online operations of traditional media, the dynamics of not one but two newsrooms, the online one and its traditional media counterpart, as well as their interactions, may shape what constitutes the news, who reports it, and when it is made available to the public. Second, advertising and marketing personnel may also influence what gets covered, via topic selection and budget allocation, to a greater extent than what is usually the case in print newspapers.[6] Third, technical and design personnel also seem to inform how the news gets reported, from the use of multimedia and interactive tools to the adoption of a notion of the visual interface as an integral part of the storytelling effort. Fourth, by voicing their opinions in forums, chat rooms, and publications housed within the new-media outlet, and hyperlinking these web pages to other sites from personal weblogs to the homepages of advocacy groups, users appear to shape what is seen as newsworthy, who gets to communicate about it, and how it gets covered.

This increase in the array of actors who shape the news in the online environment invites a shift in our understanding of the locus of news production. In his research on the construction of art, Becker (1982, p. 34) coined the expression "art world" to refer to "all the people whose activities are necessary to the production of the characteristic works which that world, and perhaps others as well, define as art." Much as art is not only the product of artists, news in the online environment may not be (to paraphrase Gieber 1964) "what newspaper people make it"; rather, it may be what emerges from "news worlds." The composition of a particular news world and the kinds of ties that bind the relevant groups of actors would vary from one setting to the other depending on what events are deemed newsworthy, who gets to report them, and using what communication means. As with the case of art worlds, both newsness and worldness "are problematic, because the work that furnishes the starting point for the investigation may be produced in a variety of cooperating networks and under a variety of definitions" (Becker 1982, pp. 36–37). Thus, seeing the production of news in the online environment as emerging from complex and dynamic news worlds enables one to question strong *a priori* notions, mostly developed to make sense of print and broadcast media, of what counts as news and news makers.

This first transformation in the production processes brings us to the second one: the heightened importance that coordination practices across these multiple groups have in the construction of news. Rather than arising mostly from exchanges between reporters and their sources, and negotiations between reporters and editors, the news online seems to be also significantly informed by the relationships among the other groups that increasingly populate the news world. This places a premium on the work that coordinates the tasks, goals, and values of the various groups that contribute to the production of news.

The increasing relevance of cross-boundary coordination also presents an analytical challenge to the traditional way of understanding news production, which has usually looked within the newsroom and studied the work relationships of members of the journalism occupation. In chapters

4–6 I began to address this challenge by examining the relational, material, and symbolic resources utilized by the actors in the making of the Technology section, Virtual Voyager, and Community Connection. This examination has shown the importance of resources that support the coordination of production activities across disparate groups, such as positions that bridge two otherwise disconnected work units, artifacts that are flexible enough to satisfy the informational needs of the various groups, and common linguistic tools that translate across the different meanings that groups attribute to means and ends of their joint action. Further research is needed to probe the value of these resources and elicit other resources utilized in other settings.

In relation to these transformations in the production process, there seem to be at least three potential effects in the content and form of news as it migrates to the online environment.

First, instead of being primarily journalist-centered, the news online appears increasingly to be also user centered. Sigal (1973, p. 37) wrote: "News is consensible: newspaper audiences, by their responses to news, actively shape its content. Yet the average reader has little impact on the consensual process." In contrast, in the online environment, users have a much greater direct effect on the news, from a qualitative leap in the intensity of their exchanges with journalists via email, to the presentation of their own views of journalist-authored stories on online papers' forums, to the publication of their own newsletter within the online paper. A trend toward more user-centered online news could *de facto* deepen the "civic" or "public" journalism movement, which has sought a greater involvement of the citizenry in the editorial process and the publication of "all the news that citizens want to know" (Charity 1995, p. 19). Furthermore, the growing influence of marketing and advertising personnel, usually sensitive to the preferences and needs of consumers, may also, directly or indirectly, add to a heightened user centeredness of news online. The aggregate effect may be an expansion in the news available to the users of a site, in terms of both events covered and perspectives adopted on any topic.

Second, instead of being fundamentally a monologue communicated unidirectionally and adding very few, if any, responses from readers in venues such as letters to the editor, the news online appears to increasingly include these unidirectional statements within a broader spectrum of ongoing conversations. That is, the online coverage of an event (especially, but not exclusively, a high-profile event) tends to elicit a wider spectrum of voices and the explicit and implicit exchanges among them.

This, in turn, opens the news to a higher degree of contestation, expressed either by direct conflict of opinions or indirect multiplicity of views, than the typical case of traditional media. The news as conversation may be partly due to journalists' increased awareness of their audience's viewpoints. It may also be partly the result of the growing authorship of new media content by members of the public, housed both within traditional news sources such as online papers and nontraditional ones such as personal weblogs. Whether or not some of this conversational content is consider as news by currently working journalists, my research provides enough grounds to suggest that it may be becoming increasingly newsworthy to the audience of new-media news. The relevance assigned to emails from their audience by contributors of the Technology section, the exchanges between sailors and vicarious stowaways in Virtual Voyager, and the nonprofit sites in Community Connection provide windows into the dynamics and value of the news as ongoing conversations.

Third, in addition to the local and national emphasis of most news reported in print and broadcast media, online news also appears to present a micro-local focus, featuring content of interest to small communities of users defined either by common interests or geographic location or both. If innovations such as Community Connection are a good indication of the implications of this micro-local focus, the news online may also feature specialized and utility-based content that differs from the more generalist orientation of most mass media content. In addition, at the geographic level, this micro-localization of the online news would expand the trend toward what is called "zoning," or the creation of specialized editions by area of distribution that many metropolitan dailies have implemented in response to the suburbanization of their readership since the 1970s.[7] News of import to micro-local audiences, from high-school sports games to the activities of narrowly specialized nonprofit organizations, rarely get featured in traditional media, even in the pages of zoned editions of print dailies. In view of the economics of online communications and the increasing role of users in the dynamics of news worlds, micro-local content may gain prominence in online news as larger segments of the population have access to online technologies and become familiarized with a media culture of content coproduction.

To bring the book to an end, I would like to return to my first day of fieldwork. After almost two hours of talking about Virtual Voyager, newspapers, and the web, David Galloway said something to which I have been returning ever since: "In print journalism . . . we wanted to go out and experience something and then come back and put it on a sheet of paper. We didn't notice the movement [or] hear the sound except in terms of something we could translate into a printed product. I think Virtual Voyager is making us open our eyes and ears to a form of journalism that we didn't need when we were print people." (interview, February 17, 1997)

There are two themes in Galloway's statement that get to the core of the accounts I have presented in this book.

First, print newspapers' pursuit of nonprint delivery options has not been simply a technical change to the people involved, but a fundamental cultural transformation. This transformation has been expressed not only in terms of material culture, the information infrastructure that underpins the gathering, processing, and transmission of the news, but also in the editorial and work domains. In Galloway's experience, making the news online has involved perception of new things or new interpretations of previously perceived things and communication of these perceptions and interpretations in a new fashion. Echoes of this theme recur in many parts of this book, such as, for instance, when Sara Glines, who became editor of New Jersey Online after a two-decade career in print journalism, commented that user-authored content in online environments had begun to alter what "editorial" meant for her. These transformations in material and communication culture have been tied to changes in the nature of work, such as challenges to the very identities of the occupations and organizations that constitute the newspaper industry. To Galloway, an occupational identity alternative to print journalism has been enacted in relation to multimedia storytelling. Issues about the identities of newspaper organizations as they have ventured into nonprint territories also appear throughout this book such as, for example, when Arthur Sulzberger Jr., publisher of the *New York Times*, reflected on the identity quandaries of his paper in its expansion from print to online.

The second theme raised by Galloway is making sense of this cultural transformation in relation to print: according to him, things were not the same "when we were print people." I have wrestled throughout this book with this popular notion of the end of print in new-media spaces. My analysis has shown that American dailies have often tried to reproduce print's ways of doing things in their nonprint forays. But in doing this they have begun constructing a kind of newspaper that although it bears connections to its print predecessor, also differs qualitatively from it in its material infrastructure, editorial practices, and production routines.

Print has survived in the online environment, but, paradoxically, this survival has enabled the creation of a new medium increasingly dissimilar from the old one. At the time of writing this book, what will ultimately result from this process in which the pursuit of sameness has led to unintended novelty remains an open question. Which is why understanding its dynamics is so critical, both to capture a sense of contingency and indeterminacy that will be much more elusive when the dust settles and to try to influence its evolution in desired directions.

Appendix
Research Design

The research for this book combined ethnographic case studies of contemporary innovation efforts by online papers with archival research about consumer-oriented electronic publishing initiatives by American daily newspapers from the videotex endeavors of the early 1980s to the web projects of the late 1990s.

Ethnographic Case Studies

Between 1997 and 1999, I conducted case studies of innovative projects in three online newspapers: the New York Times on the Web's Technology section, HoustonChronicle.com's Virtual Voyager, and New Jersey Online's Community Connection. I selected these cases because they shared some important characteristics: they involved the creation of original content on a regular basis and attempted to take advantage of some of the web's distinctive features as an information environment. In addition, these cases were also selected because of two features they did not share: they illuminated what happened to different types of traditional print paper content in their online incarnation, and they aimed to exploit different aspects of the web as an information environment. First, whereas the Technology section focused on a combination of news and opinion content, Virtual Voyager concentrated on general-interest or features material, and Community Connection worked with user-authored content. Second, whereas tinkering with immediacy and interactivity was the original intention behind the Technology section, multimedia storytelling was the defining trait of Virtual Voyager, and turning users into co-producers was the key goal of Community Connection. This combination of shared and unshared features enabled me to expect enough commonality across the cases to make sensible comparisons and enough difference to illuminate various technical, communication, and organizational alternatives.

I spent between 4 and 5 months per case. I observed the work practices of those most directly related to the three projects under study and conducted 142 interviews with relevant actors. I held many more informal conversations with interviewees and people whom I did not formally interview. Because these projects were seen primarily as editorial initiatives, they were located within the editorial units of their respective organizations. Although I focused my observations on these editorial matters, I paid special attention to their intersections with the work of advertising, design, marketing, and technical personnel, with the goal of acquiring as comprehensive an understanding as possible of the dynamics behind the developments of the projects under study. I also interviewed most of the people directly related to the three projects, as well as other relevant actors, including users, when they were directly engaged in the production of content featured in these projects. I talked with people from all the occupations listed above and with those in all the hierarchical levels of full-time employees in the different organizations.

In addition to interviews and observations, I collected a wide array of relevant public and corporate documents, such as brochures, press releases, articles, memos, and how-to guides, and examined a large sample of the sites produced by these projects during their existence.

I assumed an overt stance with the people I observed and interviewed. I explained to them the nature of my research and the types of outlets where I expected to publish its results. I also asked them how they preferred to be identified should I decide to quote any of their statements.[1] The majority chose to be identified with their real names and positions, although some asked me to attribute their statements to a generic category. Thus, statements are attributed to an individual's name and position or to the generic category of an individual's choice, such as "an editorial staff member at XYZ online newspaper." If the interview took place after the individual had ceased working at the online paper, I attributed that actor's statements to his or her latest affiliation with a relevant unit of the paper or its corporate parent.

I enjoyed a congenial environment for undertaking my research: personnel at HoustonChronicle.com, New Jersey Online, and the New York Times on the Web generously shared their time and knowledge, and provided almost unrestricted access to the different activities that compose their work routines. Once fieldwork for each case had been completed, I prepared a report with relevant preliminary findings and ideas for future developments. I wrote these reports partly to give something back to the people and projects I had studied shortly after my fieldwork had concluded. Each report was usually followed by exchanges with each respective organization about what I had written, a process that in turn contributed to my understanding of the dynamics of each case.

Archival Research

I conducted archival research of the newspaper industry's trade publications from 1969 to 1999 and complemented the findings from this original material with secondary sources. I gathered both text and graphic data. Regarding text materials, I read each story, news brief, editorial, column, op-ed article, letter to the editor, and classified advertisement that dealt with attempts to extend newspapers' franchise beyond ink on paper. Concerning visual materials, I collected each display advertisement, cartoon, and issue cover about the same matters. I focused on both the events reported and actors' reactions to these events.

For the period 1969–1993, I looked at *Editor & Publisher*, a weekly that has been the main trade publication of the American newspaper industry for more than 100 years. For the period 1994–1999, the first years of American dailies' publishing on the World Wide Web, I drew from a wider array of sources for two reasons: the contemporaneous focus of this study, and the disregarding of contextual matters in most secondary sources. Thus, in addition to *Editor & Publisher*, I also looked at the following sources:

• Mediainfo.com, a bimonthly publication of *Editor & Publisher* that focused exclusively on the online news industry, and E&P Online, *Editor & Publisher*'s weekly web site

• the *American Journalism Review*, a monthly trade publication, and its web site, AJR Newslink

• the *Columbia Journalism Review*, a bimonthly trade publication published by Columbia University's Graduate School of Journalism

• *Nieman Reports*, a quarterly edited by Harvard University's Nieman Foundation.

Except for Medianifo.com (which was launched in 1997) and E&P Online and AJR Newslink (which began featuring original content around 1996), all the sources were examined for the entire period1994–1999. In all, I analyzed approximately 1,600 issues of trade publications spanning a 30-year period.

Notes

Chapter 1

1. Print newspapers in the United States also ventured into electronic publishing for businesses. Dow Jones's News/Retrieval, and Knight-Ridder's VU/TEXT are two prominent examples of innovations in this area using videotex technology. Finn and Stewart (1985) provide an analysis of these early initiatives aimed at the business market. However, in this book, I concentrate on consumer-oriented efforts, insofar as this has been the primary publishing market for newspapers.

2. On the global homogenization of the modern newspaper form during the last decades, see Barnhurst and Nerone 2001.

3. Studies in the sociology of news production that have dealt with newspapers have illuminated different aspects of this ensemble of communication, technology, and organization (Argyris 1974; Breed 1955; Clayman and Reisner 1998; Ettema and Glasser 1998; Fishman 1980; Kaniss 1991; Lester 1974; Roshco 1975; Sigal 1973; Sigelman 1973; Sokolski 1989; Stark 1962; Tuchman 1978; Warren 1967).

4. On these early fax papers, see Hotaling 1948; Shefrin 1949.

5. For analyses of these trends, see Compaine 1980; Picard and Brody 1997; Smith 1980; Stone 1987.

6. The World Wide Web is structured on a client-server model of information exchange. 'Client' and 'server' can refer to both software and hardware. The client computer connects to a server computer in which the information is stored. Then, the client computer uses the appropriate client software to request the information from the server computer, which is processed by the server software and then delivered to the client computer. In a typical exchange, a regular user accesses the web using a personal computer as the client machine, which then requests information to a larger server machine, also known as an Internet host.

7. The weekday circulation of the print *USA Today* in the same period was 1.6 million.

8. For different perspectives on this matter, see Bijker 1995a; Bowker and Star 1999; Fujimura 1996; Jasanoff forthcoming; Kline 2000; Latour 1993; Orlikowski 2000; Pickering 1995; Yates 1993.

9. That process was first identified in the study of media artifacts by Rice and Rogers (1980).

10. Although Lievrouw and Livingstone (2002) define "new" media in this way, and online newspapers as one such new medium, this characterization also applies to "old" media.

11. I introduce conceptual resources when I first use them to analyze a particular aspect of the empirical material.

12. In general, an interface is the boundary across which two systems communicate. In this book, the term interface is employed to refer to the "graphical user interface," which is the visual configuration of a personal computer screen representing both the input and output of the information as seen by the user. That is, a personal computer user may request information to a web server by double-clicking on an icon on the computer's screen, which results in the delivery of a file also seen on the screen.

13. HTML was the dominant coding language used to build pages available on the World Wide Web during the period of this study. This language contains commands or tags that tell the user's computer how to display on the screen the different elements of a page, such as text, graphics, audio, video, animation, and links to other pages.

Chapter 2

1. See e.g. Campbell and Thomas 1981; Sommer 1983; Tydeman, Lipinski, Adler, Nyhan, and Zwimpfer 1982; Tyler 1979; Wilkinson 1980.

2. See e.g. Desbarats 1981; Marchand 1987; Mayntz and Schneider 1988; Miles 1992; Sigel 1983b; Vedel and Charon 1989.

3. See e.g. Blomquist 1985; Criner 1980; Johansen, Nyhan, and Plummer 1980; Noll 1980; Sigel 1980. One of the few consumer-oriented videotex initiatives in which the U.S. government was involved was Project Green Thumb, designed to provide news and information to farmers. For more details, see Rice and Paisley 1982.

4. For accounts of various aspects of this process, see Baer and Greenberger 1987; Endres 1985; Marvin 1980; Picard and Brody 1997; Smith 1980; Weaver and Wilhoit 1986.

5. On the notion of technological determinism, see Bimber 1990; MacKenzie 1984; Misa 1988; Scranton 1994; Smith 1994; Staudenmaier 1989; Williams and Edge 1996. On its use in the study of information and communication artifacts, see Edwards 1995; Hamilton 1997; Kling 1994; Kling and Iacono 1988; Pfaffenberger 1989; Roscoe 1999; Winner 1986.

6. "Electronic newspaper found unprofitable," *Editor & Publisher*, August 28, 1982, pp. 7–8.

7. "AP finds meager demand for electronic news," *Editor & Publisher*, October 2, 1982, pp. 10, 20.

8. "Public access videotex," *Editor & Publisher*, March 23, 1985, pp. 40–42.

9. The newspaper association argued that AT&T's size would result in unfair competition in a nascent industry and that the provision of content should be separated from the provision of the conduit over which it was transmitted. In August 1982, a court ruling prohibited AT&T and the Regional Bell Operating Companies from entering the electronic publishing field for the next seven years (Maguire 1982). For different perspectives on this matter, see Branscomb 1988; LeGates 1984; Mosco 1982; Neustadt 1982; Pool 1983.

10. In the early 1980s, the only comparable videotex initiative (in scope and size) by an American newspaper company was Times Mirror's Gateway system.

11. The TV set was disabled to receive television signals, and information was transmitted over a dedicated phone line to make use of the system as convenient as possible.

12. For a study of users' initial reactions to the interface, see Atwater, Heeter, and Brown 1985.

13. AT&T had originally priced the unit at $900, but after negotiations with Knight-Ridder it agreed to sell it for $600. Fidler (1997) has said that planning by Knight-Ridder and AT&T officials assumed that Sceptre would cost approximately $100, and that Knight-Ridder executives were dismayed to learn about the higher cost and thus the higher market price of the unit.

14. A similar strategy was pursued by other videotex newspapers, such as Times Mirror's Gateway, which were originally designed for dedicated terminals.

15. Aumente (1987, p. 4) uses the image of "an overbuilt luxury liner that could not negotiate tight turns to reach safe harbor" to describe Viewtron's and Gateway's failed attempts to reach personal computer users.

16. The system had been made available to personal computer users nationwide thanks to partnerships with local newspapers.

17. Times Mirror's Gateway had folded a few months ago, also after multi-million-dollar losses. For an account of this project, see Noll 1985.

18. Users' demand for interpersonal communication was prevalent in several other newspaper-related videotex efforts. For instance, the *Spokesman Review* and the *Spokane Chronicle* launched Electronic Editions (videotex services) in October 1983. A year later, they eliminated the news. After studying users' habits, they transformed the system into a massive bulletin board. According to Shaun Higgins, director of Electronic Editions, "We were simply spending more time

editing the news than people did reading it." (Miller 1985b, p. 27) Along the same lines, several researchers have noted the key role played by enabling and fostering user-generated content in the success of the French Minitel system, the only commercial videotex success story of the 1980s (Charon 1987; Feenberg 1995; Iwaasa 1988; Schneider, Charon, Miles, Thomas and Vedel 1991).

19. Teletext also became available to cable television subscribers. There were some technical differences involved in sending a signal over cable wire rather than over the air, but the service and their implications for newspapers were relatively similar for the analysis presented here.

20. Television frames are composed of a fixed number of lines, but a small number of them do not carry information and are only visible to the viewer as black bars at the top and bottom of the screen when the signal is poorly adjusted. These are the vertical blanking intervals.

21. On users' reactions to this experiment, see Elton and Carey 1983.

22. The network was under military control since its inception at the Department of Defense's Advanced Research Projects Agency in the 1960s and until the 1980s, when control was transferred to the National Science Foundation. This opened up its use considerably by the scientific community, but commercial exploitation was heavily restricted because the network was still funded by public resources.

23. From another angle, some scholars have argued along similar lines. Bromley and Bowles (1995, p. 23) suggested that "with a flat and aging subscriber base and decline in percentage of advertising dollars, the industry can ill-afford a wait-and-see attitude or to wait for technological perfection." Brill (1999, p. 163) stated that "as a publishing opportunity, the newspaper industry cannot afford to ignore this new medium."

24. For an account of a European initiative aimed at building a news-oriented portable digital assistant, see Molina 1999.

25. StarText was phased out in 1997, a year after the launch of the *Star Telegram*'s web site.

26. Belo Corp., Central Newspapers Inc., Cowles Media Co., Freedom Communications Inc., McClatchy Newspapers, and Pulitzer Publishing Co.

27. Interestingly, though not surprisingly, most of these early web developments took place in "high-tech" enclaves. The first student newspaper was published at MIT (Garneau 1995); the *Palo Alto Weekly* is supposed to be the first for-profit paper (Carlson 2000); the first major site related to a newspaper was NandO, affiliated with the *News and Observer* of Raleigh, North Carolina, in the heart of the so-called Research Triangle.

28. See e.g. Carveth, Owers, and Alexander 1998; Garrison 1997; Martin and Hansen 1998.

29. See e.g. Bijker 1995b; MacKenzie and Wajcman 1985b; Staudenmaier 1989; Tushman and Rosenkopf 1992; Williams and Edge 1996.

30. See e.g. Bijker 1995a; Elzen 1986; Garud and Rappa 1994; Kline and Pinch 1996; Pinch 1996; Pinch and Bijker 1984.

31. This seems, at least partially, to be an artifact of the strong influence that the empirical program of relativism in the sociology of science (Collins 1975, 1981, 1992) had in the foundation of the model (Boczkowski 1996). Much as the former focused on controversies in science, the latter initially concentrated on controversies in technology. Once the controversy under examination was over, the analysis ended. Thus, there was not much incentive to examine what happens after closure is achieved. For a recent attempt to fill this void in the sociology of science, see Simon 1999.

32. Misa (1992) has argued for the need to see closure not as in opposition to, but as a pre-condition of, change: without a certain level of stability it is difficult for something to evolve.

33. For descriptions of closure mechanisms, see Kline and Pinch 1996; Misa 1992; Pinch and Bijker 1984.

34. See e.g. Cockburn and Ormond 1993; Fischer 1992; Kline and Pinch 1996; Suchman 2000; Tyre and Orlikowski 1994.

35. See e.g. Mackay and Gillespie 1992; Pinch and Trocco 2002; Rosen 1993.

36. This relates to a phenomenon studied under the guise of "network externalities" by economists (Katz and Shapiro 1985) and under the guise of "critical mass" by communication scholars (Markus 1987): under certain conditions, the development of a new technology is accelerated if the benefit accrued to each user grows with its market penetration. For instance, the benefit that an individual derives from being connected to a telephone system grows with each new user of the system. Research on "increasing returns to adoption" (Arthur 1988) and "path dependence" (David 1986) has also highlighted the role of network effects in technological development.

37. Of course, this is not the first time an established communication technology business innovates defensively. The history of AT&T and radio resonates with the case of newspapers and videotex and has a twist in view of papers' efforts to keep the phone company out of electronic publishing in the 1980s. Seeking to secure its position in wired-based communications, in the early 1920s AT&T participated in a cross-licensing agreement with General Electric, Radio Corporation of America, and Westinghouse, which guaranteed it exclusive licenses in wired telephony and telegraphy. AT&T did this under the assumption that the future of radio would be tied to its existing network. However, soon after the agreement broadcasting emerged as the preferred option for consumers, and profits seemed to be in manufacturing reception and transmission units. Thus, the reaction of AT&T executives "was defensive in nature. Using patent rights granted by the agreement to a monopoly on all activity related to wired

telephony, they prohibited members of the GE group from using telephone lines to send signals from remote pickups . . . to their broadcasting studies" (Reich 1985, p. 226). Later, AT&T attempted other strategies aimed at taking advantage of its telephony networks, but they were not very successful. Due to an adverse legal ruling, AT&T retreated to telephony service a few years later.

Chapter 3

1. Popularization of the web opened up a new market for syndicated services, which saw non-newspaper sites becoming heavily interested in their products.

2. During the second half of the 1990s, the online news industry used three notions to count traffic of web sites: the *hit*, the *page view*, and the *unique user or visitor*. A hit is a single request made by a client computer for a file residing in a server computer. A web page may contain one or more files. For instance, a web page may contain one text file and three graphical files that indicate the option to go to the previous, the next, or the home page. When the client computer requests this page from the server, this would amount to four hits. The page view, used to facilitate comparison across sites that may construct web pages using very different numbers of files, refers to the request for a single page by a client computer to a server computer, disregarding the number of files contained in that page. Whereas a hit and a page view measure usage of a site's files, "unique user" measures the number of client computers that access the site. A site's server records the information about the IP address of the client computer and counts the number of times this computer requests information to the site during a given period, usually a day or a month.

3. In 1998, according to its web site, Knight-Ridder "dropped the hyphen."

4. In their longitudinal study of three Scandinavian online newspapers in the period 1996–1999, Eriksen and Ilström (1999) report a similar shift.

5. See e.g. Eriksen and Sørgaard 1996; Light 1999; Massey and Levy 1999; Neuberger, Tonnemacher, Biebl and Duck 1998; Palmer and Eriksen 1999.

6. My adoption of 'recombination' was inspired by Stark's (1996) use of that word to address the cultural transformations of transitional economies in Eastern Europe, and by Henderson and Clark's (1990) idea of "architectural innovation." The focus of this subsection is more descriptive and less analytically ambitious than Stark's and Henderson and Clark's. It is also worth noticing that Lievrouw, Bucy, Finn, Frindte, Gershon, Haythornthwaite, Köhler, Metz, and Sundar (2000) have adopted the idea of recombination in a recent review of the literature on new media.

7. On the Media Lab's work on news customization in the 1980s, see Brand 1987. On "FishWrap," see Bender, Chesnais, Elo, Shaw and Shaw 1996. For opposing views on the customization of editorial content, see Negroponte 1996 and Sunstein 2001.

8. On the history, dynamics, and implications of advertisement targeting in different media, see Turow 1997.

9. Horizontal in the sense of lacking the depth of being able to get more information on a story than what is printed in each edition. In computer parlance, print presents a "what you see is what you get" interface; in contrast, dynamic web pages feature a "what you search is what you get" interface.

10. Advance Publications, the Donrey Media Group, the E. W. Scripps Company, the Hearst Corporation, and the MediaNews Group.

11. In 1996, classifieds accounted for about $15 billion, nearly 40% of American dailies' advertising revenues (Newspaper Association of America 1998).

12. Peng, Them, and Xiaoming (1999) found that 80% of their respondents' sites had such archives.

13. See also Singer, Tharp, and Haruta 1999.

14. On the role of updates in online publishing, see Aronson, Sylvie, and Todd 1996; Cameron, Curtin, Hollander, Nowak, and Schamp 1996; Eriksen and Ihlström 1999; Neuberger, Tonnemacher, Biebl, and Duck 1998.

15. For an account of the making of this special, see Borum 1998, pp. 75–79.

16. Thus, this project also illustrates the potential of the "furnace" model of news production: one large multidisciplinary team generating content for availability in print, broadcast, and online environments (Harper 1998).

17. A somewhat similar situation took place in broadcast radio's formative years. Douglas (1987, p. 197) has shown that "in the hands of amateurs, all sorts of technical recycling and adaptive reuse took place." Thus, diverse combinations of tools and skills meant very different listening experiences, including variations in what could be listened to and with what quality. This matter remained present through the early years of radio broadcasting, to the point that Smulyan (1994, p. 19) has referred to users as "listeners/technicians": "Listeners reached a technological limit in their ability to pick up distant stations and found that stations had an annoying tendency to fade away, despite the best efforts of the listeners/technicians."

18. For analyses of different facets of user authorship in online news, see Friedland 1996; Kenney, Gorelik and Mwangi 2000; King 1998; Light and Rogers 1999; Pride 1998; Schultz 1999, 2000.

19. I will discuss the issue of user-authored content in print and online papers at greater length in chapter 6.

20. The search for a successful economic model for videotex newspapers and, especially, online newspapers resonates with the commercialization of broadcast radio. According to Smulyan (1994, p. 39): "The question of 'who pays for broadcasting?' occurred repeatedly in article and chapter titles of the early 1920s, and

in the recollections of observers. Early broadcast radio had huge financial problems. A variety of different businesses from feed stores to newspapers founded and financed radio stations but provided them with only small budgets. Faced with the need for day-in and day-out programming, broadcasters relied first on amateur musicians and then on those professional performers they could convince to appear for free. As listeners began to turn away from long-distance listening and to seek improved programming, broadcasters faced increasing pressure and anxiety." By the early 1930s, a sophisticated ensemble of technical, programming, and policy decisions had generated the advertising-supported, mass-communicated product familiar to us today. "The commercialized network system had succeeded so well that earlier confusion surrounding the shape, content, and financing of broadcasting was forgotten" (ibid., p. 165). Perhaps the enigma of online newspapers' business model will also take a decade or so to be solved. If that were the case, this chapter should serve as an antidote to the deterministic and naturalizing temptations informing such "forgetting of earlier confusions" by highlighting the contingent character of the path to commercial success, something always popular when it comes to understanding new communication technologies.

21. On the coexistence of different funding models in interactive media, see McMillan 1998.

22. Among the online newspapers of large corporations, two exceptions to this trend were USA Today, which had become profitable in September 1998 (although no information was given about the financial and accounting details behind that statement) and the Wall Street Journal Interactive Edition (the only major player using a subscription-only model, which was expected to turn a profit in 1999—see Neuwirth 1998b).

23. See e.g. Kogut, Shan and Walker 1992; Lane and Mansfield 1996; Padgett and Ansell 1993; Powell 1996; Sabel and Zeitlin 1997; Stark 1996.

24. Many changes have indeed take place in the print environment, i.e., the introduction of illustrations, photographs, and color, but always resulting from the combination of ink and paper.

Chapter 4

1. See Hutchins 1995.

2. All interviews cited herein were conducted by the author.

3. As we saw in chapter 3, specials consisted of an in-depth look at a phenomenon or matter of particular attractiveness, from a major sports event to a salient health-care issue. For instance, in October 1996 CyberTimes contributor Peter Wayner put together a special, "Computer Simulations: New-Media Tools for Online Journalism," that looked at the epidemiological dynamics of HIV. Instead of merely describing it, he also developed a simulator that allowed the user to tin-

ker with the behavior of different factors that influence the spread of the virus in the population. The online environment, according to him, provides "a new opportunity to distribute not just words but simulated worlds that let the reader experiment with and control a tiny environment." Thus, "online journalists are now able to make a point by using a computer simulation to illustrate how the reader's own adjustments or a model can predict some aspect of the real world" (Wayner 1996).

4. The columns were arts @ large (on the digital arts scene), CyberLaw Journal (on the legal implications on computer and communication tools), Education (on the effects of technology on teaching and learning), Eurobytes (on cyber issues in Europe), and TraveLog (on mobile computing).

5. See e.g. Altheide 1976; Epstein 1973; Gans 1980; Gitlin 1980; Kaniss 1991; McManus 1994; Tuchman 1978.

6. Because this is a study of online newspapers on the web, in this and the following sections I will concentrate on the work involved in creating material primarily intended for CyberTimes. That is, I will not delve into the reporting practices that lead to content written primarily for the print *Times*.

7. There were about a dozen "stable" contributors.

8. Some of my interviewees said that one difference was the extended use of email and the web to gather information. Although this is different from reporting practices before email became a popular communication tool, there is no indication that such use of email and the web is an exclusive feature of online journalism. On the contrary, anecdotal evidence suggests that this practice is becoming increasingly mainstream in print newsrooms. In view of the history of twentieth-century journalism, this should not be surprising: as the diffusion of the telephone was related to some substitution of "leg work" by "phone work," the mainstreaming of email and the web could very well lead to a substitution of them by "online work."

9. Between November 12, 1997 and June 12, 1998.

10. Mirror servers are located inside a "firewall." Files saved in them are not accessible to the public. When the files are ready for public consumption, they are moved to servers that can be accessed by regular users.

11. A "drop cap" format option turned an article's initial letter, in the body of the text, not in the title, into an uppercase letter with a larger font than the rest of the text. The "bio of the reporter" option put a hyperlink to a page containing a brief biographical sketch of that story's author. The "internal link" option placed a hyperlink to another page of the Times on the Web, but different from the bio of the reporter. The "external link" option placed a hyperlink to a file located on a web site other than the Times on the Web.

12. The adjustment of the temporal pattern of one activity to synchronize with that of another is a process known as "entrainment" (McGrath and Kelly 1986).

According to Ancona and Chong (1996, p. 273), entrainment has an array of consequences in the dynamics of work. On the positive side, visible in the present case, it offers workers dealing with complex sets of tasks the possibility of "coordination by time rather than by activity." On the negative side, also prevalent in the Technology section, "entrainment can hamper creativity [because] it represents repeated patterns of activity, hence by its very definition there is an emphasis on repetition, not innovation" (ibid., p. 278).

13. There were exceptions. For example, the wire feed from news agencies was automatically updated every time a new relevant news item was available, and some breaking technology news, in which case stories were beginning to be published soon after events had occurred, as part of a general push within the whole site to provide users with continual updates.

14. The actors used the print-publishing term "front page" rather than the web-creole term "home page."

15. A plug-in is an accessory program that enhances a main application. Multimedia and animation plug-ins are often used to enhance the textual capabilities of web browsers.

16. See e.g. Akrich 1995; Bardini and Horvath 1995; Bijker 1995a; Carlson 1992; Orlikowski and Gash 1994; Woolgar 1991.

17. See Mackay, Carne, Beynon-Davies, and Tudhope 2000.

18. Contributors did not usually share these exchanges or a summary of them with their editors, thus partly diminishing the direct impact that this type of communication practices could have had upon the section as a whole.

19. The history of radio broadcasting provides another interesting parallel here. In a practice that continued well into the 1930s, announcers usually asked listeners to send letters to provide a sense of who was listening, from where, and what programs people liked. According to Smulyan (1994, p. 96), "listeners' letters contained suggestions about every aspect of broadcasting, from programming to hours on the air."

20. See Boczkowski 1999b.

21. In mid 1998 the Times on the Web had more than 150 forums organized around various topics. According to estimations by Justin Peacock and Cynthia Toletino, the two people in the editorial department who managed the forums, by August 1998 the forums were contributing an aggregate of 2,000–2,500 new posts per day (interview, August 14, 1998).

22. In a study of the *Times*'s print reporters, Schultz (2000, p. 214) also found a lack of interest in the web edition's forums: "From the surveyed . . . journalists, 12 out of 19 admitted that they do not even visit the Times's own online forums. Only six claimed to visit the discussion sites 'from time to time'. No one visited them regularly."

23. These events included Bill Gates's deposition before the Senate Judiciary Committee on March 4, a proposal made by a group of computer industry executives to the Department of Justice with a set of remedies to curtail Microsoft's alleged monopoly power on April 7 and an aggressive public relations campaigned launched by Microsoft the day after, and the filing of an antitrust suit against Microsoft by the federal government and 20 state attorneys on May 19.

24. The lack of mention of an article in a message does not rule out its possible influence over the content of what was posted. That, however, was impossible to determine with the information available. Moreover, it could also be argued that what people say in forums is generally influenced by what appears in the media. However, this may also be true in any other exchange of information over a topic covered by the press and does not exclude the relevance of the issue about the explicit and direct relationship between stories and messages as two sources of content included by newspaper-related web sites.

25. Anecdotal evidence suggests that, more often than suspected, users—sometimes in large groups—migrate from one site to another offering what they perceive to be as a better discussion context.

26. See e.g. Borum 1998; Endres 1998; Huxford 2000; Martin and Hansen 1998.

27. See e.g. Brill 1999; Harper 1998; Pavlik 1998; Williams 1998.

28. See e.g. Friedland 1996; Jankowski and van Selm 2000; King 1998; Light and Rogers 1999; Schultz 2000.

29. See e.g. Bendifallah and Scacchi 1987; Fujimura 1996; Gasser 1986; Schmidt and Bannon 1992; Star 1991; Strauss 1985; Suchman 1996.

30. The use of the notion of articulation work to address issues of alignment between different work units is informed by Fujimura's (1987) study of these issues in the context of scientific practice.

31. See e.g. Burt 1992; DiMaggio 1992; Granovetter 1973; Hargadon and Sutton 1997; Marsden 1982; Padgett and Ansell 1993.

Chapter 5

1. The department was called "Content," instead of the usual term "Editorial," to signal differences between print and online publishing environments.

2. Hearst also established a New Media Center—run by Hearst New Media and Technology in the corporation's Manhattan headquarters—to carry on training, research, and development tasks related to its units' online efforts.

3. In 1998 the *Chronicle* averaged more than 550,000 in daily circulation and 750,000 on Sunday.

4. Before that restructuring, Evangelista, Galloway, and Golightly had other responsibilities in the online operation in addition to producing voyages.

5. Earlier voyages had been undertaken without any fixed frequency.

6. The electronic products division leased the special camera with a pre-specified amount of shots available from its manufacturer, IPIX Corporation.

7. That is how IPIX technology for 360° photography worked: the photographer positioned the camera on a tripod, shot one picture, then turned 180° and shot another picture, and finally "stitched" the two halves together digitally.

8. When I conducted fieldwork in early 1998, most regular modems connecting users' computers to the Internet via regular phone lines came in four speeds of access: 14.4, 18.8, 33.6, and 56 kilobytes per second. Other options included ISDN, which stands for Integrated Services Digital Network, which allowed data to be downloaded at 128 kilobytes per second.

9. On "unruly technology," see Wynne 1988.

10. The idea was that the band and its entourage arrived in the next town sometime before noon, so that the bus and truck drivers had enough time to rest that afternoon before having to drive again at night.

11. An emulator is software or hardware that behaves like some other piece of software or hardware. For instance, Xterm allows personal computers to behave like Unix machines.

12. If there were any pictures, audio, and video, he edited and produced them following the same steps described for the "Asleep at the Wheel" voyage.

13. An old concept in the sociology of knowledge, "reification" refers to the "apprehension of human phenomena as if they were things [or] something else than human products" (Berger and Luckmann 1966, p. 89). "De-reification" points to the process that turns previously reified products into contingent outcomes of situated action.

14. See e.g. Epstein 1973; Gans 1980; Kaniss 1991; McManus 1994; Sigal 1973.

15. Needless to say, a user who did not know how to download the streaming video plug-in could have no visual experience at all.

16. See e.g. Kawamoto 1998; Li 1998; Morris and Ogan 1996; Newhagen and Levy 1998; Pavlik 1998.

17. Perl (practical extraction and reporting language) is a programming language commonly used to write CGI applications.

18. CGI (common gateway interface) is a communication protocol that enables a web server to communicate with other applications. GGI applications or scripts are most often used to allow users access to databases.

19. During the period of this study, JavaScript was one of the most popular programming tools on the web. It allows the creation of such things as interactive forms and navigational aids, which can be used without having to download plug-ins.

20. An applet is an application that uses limited memory and usually works across operating systems.

21. On the notion of translation, see Callon 1986a; Latour 1994; Law 1987; Law and Callon 1988.

22. On boundary objects, see Bowker and Star 1999; Carlile 2002; Henderson 1999; Star and Griesemer 1989.

23. For recent work within this tradition, see Landow 1997; Manovich 2001; Murray 1999; O'Donnell 1998; Poster 2001.

Chapter 6

1. This practice was not unique during the period. For instance, the *Post Boy*, published in London in the early eighteenth century, also "offered their readers a blank last page on which to inscribe their own news for onward transmission to friends in the country" (Smith 1979, p. 56).

2. For an elaboration of Thompson's ideas in the context of new media, see Slevin 2000.

3. As of March 1999, Advance Internet had eight newspaper-related sites in addition to New Jersey Online: Alabama Live, Cleveland Live, Mass Live, Michigan Live, New Orleans Live, Gulf Live, Syracuse Online, and Staten Island Live. Advance Internet also owned Rain or Shine (a weather site), The Yuckiest Site on the Internet (a children's site), Journal Square Interactive (a web development site), and Advance Internet (its own corporate site).

4. The strongly local use of these technologies of global reach resonate with Hampton and Wellman's (2002) claims about the process of "glocalization," or the mixing of global and local connectivity in contemporary computer-mediated communication.

5. Tripod and Geocities were two of the most successful self-publishing services at that time.

6. The more detailed site-building process was the following. Contributors began by going through a registration process in which they provided basic data about their organizations that later were fed into a searchable database of Community Connection sites. Once that information had been submitted, users were presented with the New Jersey Online Community Home Pages Agreement, which regulated the relationship between New Jersey Online and Community Connection contributors. This agreement stated that the use of Community Connection was free of charge, that all New Jersey not-for-profit groups were eligible, that the information contained in the groups' sites should not be in any way

harmful, and several precautions concerning privacy matters, among other issues. Once they accepted the conditions stated in the agreement, contributors continued building their sites by selecting an image for their homepages. Sites could contain one image, and contributors could either choose from an New Jersey Online library of photographs or upload their own. This was probably the most difficult step technically. To give an idea of the level of technical expertise involved, I include the full set of instructions given to upload an image: "• Save the desired image on your computer as a GIF or JPEG format. • Click **Browse. . .** a window will open. • Find an image file on your computer that you want to put on your site. Click on it twice. • The image file you selected will be entered into the **My Image** box. • Click **SUBMIT** and preview your image. **Tip**: For best results, choose an image that is squarish in size, rather than long and narrow. The actual size of the image that will appear on your site's Home Page is 150 pixels wide by 110 pixels high. **Important Note**: You must have the rights to use any image that appears on your group's site. Do not use a copyrighted image without permission. **Internet Explorer Users**: You must have version 4.0 or higher to use this feature." Once the image issue had been solved—choosing one from New Jersey Online's library, uploading one's own, or not including any—the user was asked to complete the "About Us/Home Page section" of the group's site. This section was the only one that had to be completed for a site to be approved and published within New Jersey Online. All other sections were optional, and some sites—especially those belonging to groups that already had a site and just wanted to use Community Connection to link to it—only built this page. In addition to the "About Us/Home Page," six other sections could be included in the site. The remaining sections were: "News," "Important Dates," "Who's Who," "Get Involved," "Congrats & Thanks," and "Links." The first five sections had a 500-word limit, and only five URLs could be included in the last section. When contributors updated the News section, past content could be archived to give a sense of continuity.

7. For different perspectives on the normalizing effect of editorial conventions in journalism, see Carey 1986; Darnton 1975; Eliasoph 1988; Gitlin 1980; Schudson 1982.

8. Some authors have indicated that the presence on stories of email links to reporters is related to a substantial growth in contact from users (Boczkowski 1999b; Harper 1998; Patrick, Black and Whalen 1996; Riley, Keough, Christiansen, Meilich, and Pierson 1998).

9. Although it could be argued that readers of print papers also tend to read only parts of it, the whole artifact is easily available to them, and its total consumption is possible in a relatively limited amount of time. The same cannot be said of more than 3,000 sites.

10. In this sense, the distinction between generalized and specialized goods gets blurred in the online environment even more sharply than in other industries in which computer technologies have been used to flexibilize the production process. On this latter point, see Coriat 1997; Ittner and Kogut 1995; Pine 1993; Piore and Sabel 1984; Streeck 1991.

11. See e.g. Bromley and Turner 1997; Kenney, Gorelik, and Mwangi 2000; Massey and Levy 1999; Schultz 1999; Tankard and Ban 1998.

12. No data on usage patterns were made public or available to me.

13. If the publishing tool was the face the server's database presented to Community Connection contributors, the administration tool was the face it presented to New Jersey Online administrators. The tool also was an easy-to-use program, with a very simple interface consisting of fields containing information such as each entry's group name, URL, user name, password, date of creation and of last modification, approval status (yes or no), and a search engine that allowed the administrator to rapidly locate existing sites using incomplete data.

14. Whenever she needed to screen sites for approval, Alford opened the administration tool and called up the unapproved sites submitted since the last time she undertook this procedure. Then, she previewed the content of each new site, making sure that it was compliant with the guidelines stated in the user agreement. Sometimes the answer was fairly straightforward as in the case of a local chapter of a recognized nationwide charity. On other occasions, things were less evident and required Alford to conduct a careful examination of the submission. In the case of entries belonging to groups who already had a site and wanted to have a link off Community Connection, the approval process usually involved visiting the "original" site and screening it thoroughly. If a site was approved, Alford entered that information into the administration tool and submitted the changes. Depending on how busy the server was that particular day, it could take anywhere from a few minutes to several hours for the newest sites to be "live." Once a site had been approved, the procedure ended by notifying the group and welcoming it to Community Connection. If a site was not approved, because, for instance, the group was not based in New Jersey, Alford wrote an email to the group's contact person explaining the reasons for rejection. If Alford could not decide whether a site was eligible, most commonly when the nonprofit status of the group was unclear from the information provided in the submission, she wrote back to the group's representative asking for additional information.

15. The reasons for incomplete sites were very diverse, ranging from technical difficulties, to unexpected interruptions, to a realization that some required information was not available. When a site was left incomplete, Alford sent an email to the group's representative saying that the site could not be approved as submitted, and asking whether she could help in any way with the matter at hand. If she did not get a response back from the group or if the site was not completed, the Friday of that week—that choice of day was entirely arbitrary—Alford sent a reminder to the group. In the absence of any communication or change in the site's status, Alford sent a second reminder the following Friday, this time adding that the site would be deleted in a week if it were not completed by then. Seven days later, Alford proceeded to delete all the sites left incomplete that had not requested extra time to include the additional information. The reason for deleting incomplete sites was to keep the database as "clean" as possible.

16. For a summary of research on editorial gatekeeping, see Shoemaker 1991.

17. See e.g. Braverman 1974; Hirschhorn 1984; Noble 1984; Shaiken 1984; Zuboff 1988.

18. For different reflections on the relationships between editorial and commercial functions in online publishing, see Borum 1998; Chyi and Sylvie 1998; Harper 1998; Huxford 2000; Riley, Keough, Christiansen, Meilich, and Pierson 1998; Williams 1998.

19. It is worth noting the difference between this practice and what is stated in the Community Connection's user agreement: "New Jersey Online reserves the right to post advertising in the form of banners or otherwise on the Home Pages, the content, location, size and rotation frequency of which shall be determined in New Jersey Online's sole discretion. You will not be entitled to receive any revenues collected by New Jersey Online for such advertising banners." (New Jersey Online Community Home Pages Agreement, http://community.nj.com/ccregister) This discrepancy indicated that even though contributors granted New Jersey Online the right to place any type of advertisement message on their Community Connection sites, New Jersey Online officials saw that the potential gains to be derived from doing so could possibly be lost owing to the different character of the parties involved—for-profit and not-for-profit organizations—and the need for a strong cooperative ethos in the relationships among them.

20. During Community Connection's first 6 months of operation, New Jersey Online managers were unable to secure that sponsorship, although by the end of my fieldwork, negotiations with potential candidates were getting closer to a deal.

21. For diverse perspectives on newspapers' organizational structure, see Breed 1955; DuBick 1978; Esser 1998; Sigelman 1973; Sokolski 1989.

22. For related characterizations of these new organizational forms that are "neither markets nor hierarchies" (Powell 1990), see Bradach and Eccles 1989; Harrison 1994; Podolny and Page 1998; Powell 1996; Sabel 1991; Williamson 1991.

23. See e.g. Clark 1997; Cole 1996; Engeström, Miettinen, and Punamäki 1999; Hutchins 1995; Salomon 1993.

24. See e.g. Braverman 1974; Bendix 1956; Edwards 1979; Jacoby 1985.

25. See e.g. Bromley and Turner 1997; Kenney, Gorelik, and Mwangi 2000; Massey and Levy 1999; Schultz 1999; Tankard and Ban 1998. See also Friedland 1996; Jankowski and van Selm 2000; Pride 1998; Schultz 2000.

26. See e.g. Dunlop and Kling 1991; Jones 1995, 1997; Sproull and Kiesler 1991; Turkle 1995.

27. See e.g. Cummings, Sproull, and Kiesler 2002; Hampton and Wellman 1999; Matei and Ball-Rokeach 2001; Miller and Slater 2000.

Chapter 7

1. It is also worth noting that, as we saw in the case of the *New York Times*, the more extensive articulation of alignment between print and online newsrooms had transformative effects in the former that were absent in the *Houston Chronicle* and New Jersey Online cases.

2. This analysis does not exhaust the list of relevant factors shaping innovation in the newsrooms of online newspapers, since other such factors could be identified from studies of other innovation efforts.

3. One of the earliest treatments of media convergence is Pool 1983.

4. See e.g. Baldwin, McVoy, and Steinfield 1996; Black 2001; Chandler and Cortada 2000; Gillett and Vogelsang 1998; Hall 2001; Manovich 2001; Poster 2001; Schiller 1999; van Cuilenburg and Verhoest 1998.

5. See e.g. Flichy 1995; Higgins 2000; Jenkins 2001; Preston and Kerr 2002; Zavoina and Reichert 2000.

6. This could constitute a deepening of the "market-driven journalism" (McManus 1994) trend that has been growing in American media since the 1980s.

7. According to Kaniss (1991, pp. 59–60), "by the 1970s . . . it became painfully obvious to the top management at many newspapers that suburban penetration levels were dropping and that changes were necessary to win back the suburban reader. While the need for greater coverage of the suburbs was clear, this was no simple task since each region was composed of a multitude of political jurisdictions, each with its own form of government, school district, and local zoning disputes. The solution many newspapers found to deal with the political fragmentation was to fragment the newspaper itself—a process that has become known as 'zoning.'"

Appendix

I adopted a different practice in my interviews with the representatives of 31 groups participating in Community Connection. In this case, I told my interviewees that I would not associate any of their statements with their names or their groups' names, and that I would not disclose any information that could identify their groups. I based this decision on two factors. First, in light of how easy it was to create electronic mail addresses containing false names, I did not have any fast and reliable way of checking that the names given to me corresponded to either an existent person or the person effectively communicating with me. Second, the mediated and isolated nature of our contacts did not give both the interviewees and myself either a duration and directness or a referral network to build a relationship conducive to identification on a first-person basis. Based on these two gactors I judged that a default condition of anonymity was the best decision. None of the interviewees objected to this choice, although some of them expressed that they did not mind being quoted with their real names.

Bibliography

Bibliographic data on anonymous primary sources are given in the text.

Abbate, J. 1999. *Inventing the Internet.* MIT Press.

Abbott, S. 1982. How an electronic newspaper project was launched in Worcester, Mass. *Editor & Publisher,* January 23: 28–29.

Abernathy, W., and Utterback, J. 1978. Patterns of industrial innovation. *Technology Review,* June-July: 59–64.

Ahlhauser, J., ed. 1981. *Electronic Home News Delivery: Journalistic and Public Policy Implications.* School of Journalism and Center for New Communication, Indiana University.

Akrich, M. 1992. The de-scription of technical objects. In *Shaping Technology/Building Society,* ed. W. Bijker and J. Law. MIT Press.

Akrich, M. 1995. User representations: Practices, methods and sociology. In *Managing Technology in Society,* ed. A. Rip, T. Misa, and J. Schot. Pinter.

Alber, A. 1985. *Videotex/Teletext: Principles and Practices.* McGraw-Hill.

Altheide, D. 1976. *Creating Reality: How TV News Distorts Events.* Sage.

Amin, A. 1991. Flexible specialization and small firms in Italy: Myths and realities. In *Farewell to Flexibility,* ed. A. Pollert. Blackwell.

Ancona, D., and Chong, C.-L. 1996. Entrainment: Pace, cycle, and rhythm in organizational behavior. *Research in Organizational Behavior* 18: 251–284.

Anderson, H. 1997. Freelance rights online. *Editor & Publisher,* May 3: 53, 59.

Argyris, C. 1974. *Behind the Front Page: Organizational Self-Renewal in a Metropolitan Newspaper.* Jossey-Bass.

Aronson, K., Sylvie, G., and Todd, R. 1996. Real-time journalism. *Newspaper Research Journal* 17, no. 3/4: 53–67.

Arthur, B. 1988. Competing technologies: An overview. In *Technical Change and Economic Theory*, ed. G. Dosi et al. Pinter.

Ashe, R. 1991. The human element: Electronic networks succeed with relationships, not information. *The Quill*, September: 13–14.

Astor, D. 1995. United starts web site on the Internet. *Editor & Publisher*, April 8: 28.

Astor, D. 1997a. Online sales on the rise for syndicates. *Editor & Publisher*, November 15: 44–46.

Astor, D. 1997b. They set their sites on the Web in 1996. *Editor & Publisher*, January 4: 64–65.

Atwater, T., Heeter, C., and Brown, N. 1985. Foreshadowing the electronic publishing age: First exposures to Viewtron. *Journalism Quarterly* 62: 807–815.

Aumente, J. 1987. *New Electronic Pathways: Videotex, Teletext and Online Databases.* Sage.

Baer, W., and Greenberger, M. 1987. Consumer electronic publishing in the competitive environment. *Journal of Communication* 37: 49–63.

Baldwin, T., McVoy, S., and Steinfield, C. 1996. *Convergence: Integrating Media, Information and Communication.* Sage.

Bardini, T., and Horvath, A. 1995. The social construction of the personal computer user. *Journal of Communication* 45, no. 3: 40–65.

Barnhurst, K., and Nerone, J. 2001. *The Form of News: A History.* Guilford.

Batten, J. 1981. A history of K-R's Viewdata project. *Editor & Publisher*, July 4: 18, 20.

Beamish, R. 1997. The local newspaper in the age of multimedia. In *Making Local News*, ed. B. Franklin and D. Murphy. Routledge.

Becker, H. 1982. *Art Worlds.* University of California Press.

Bender, W., Chesnais, P., Elo, S., Shaw, A., and Shaw, M. 1996. Enriching communities: Harbingers of news in the future. *IBM Systems Journal* 35: 369–380.

Bendifallah, S., and Scacchi, W. 1987. Understanding software maintenance work. *IEEE Transactions on Software Engineering* SE-13, no. 3: 311–323.

Bendix, R. 1956. *Work and Authority in Industry: Ideologies of Management in the Course of Industrialization.* Wiley.

Berger, P., and Luckmann, T. 1966. *The Social Construction of Reality: A Treatise in the Sociology of Knowledge.* Doubleday.

Bijker, W. 1995a. *Of Bicycles, Bakelites, and Bulbs: Toward a Theory of Sociotechnical Change.* MIT Press.

Bijker, W. 1995b. Sociohistorical technology studies. In *Handbook of Science and Technology Studies*, ed. S. Jasanoff et al. Sage.

Bijker, W. 2001. Social Construction of Technology. In *International Encyclopedia of the Social & Behavioral Sciences*, volume 23, ed. N. Smelser and P. Baltes. Elsevier.

Bijker, W., and Bijsterveld, K. 2000. Women walking through plans: Technology, democracy, and gender identity. *Technology & Culture* 41: 485–515.

Bimber, B. 1990. Karl Marx and the three faces of technological determinism. *Social Studies of Science* 20: 333–351.

Black, D. 2001. Internet radio: A case study in medium specificity. *Media, Culture & Society* 23: 397–408.

Blomquist, D. 1985. Videotex and American politics: The more things change. . . . *Information and Behavior* 1: 406–427.

Blondheim, M. 1994. *News over the Wires: The Telegraph and the Flow of Public Information in America, 1844–1897*. Harvard University Press.

Boczkowski, P. 1996. Acerca de las relaciones entre la(s) sociología(s) de la ciencia y de la tecnología: Pasos hacia una dinámica de beneficio mutuo [On the relationships between the sociology(ies) of science and of technology: Steps towards a dynamics of mutual benefit]. *REDES, Revista de Estudios Sociales de la Ciencia* 3, no. 8: 199–227.

Boczkowski, P. 1999a. Mutual shaping of users and technologies in a national virtual community. *Journal of Communication* 49: 86–108.

Boczkowski, P. 1999b. Understanding the development of online newspapers: Using computer-mediated communication theorizing to study Internet publishing. *New Media & Society* 1: 101–126.

Boczkowski, P. 2002. The development and use of online newspapers: What research tells us and what else we might want to know. In *The Handbook of New Media*, ed. L. Lievrouw and S. Livingstone. Sage.

Bolter, J. D., and Grusin, R. 2000. *Remediation: Understanding New Media*. MIT Press.

Borrell, G. 1995. Ready, fire, arm. *Editor & Publisher*, February 4: 20TC-21TC, 26TC.

Borum, C. 1998. Navigating the Changing Landscape of News in the Information Age: Characteristics, Trends and Issues of Online Journalism. Master's thesis, University of Pennsylvania.

Bowker, G., and Star, S. 1999. *Sorting Things Out: Classification and Its Consequences*. MIT Press.

Bradach, J., and Eccles, R. 1989. Price, authority, and trust: From ideal types to plural forms. *Annual Review of Sociology* 15: 87–118.

Branscomb, A. 1988. Videotex: Global progress and comparative politics. *Journal of Communication* 38: 50–59.

Braverman, H. 1974. *Labor and Monopoly Capital.* Monthly Review Press.

Breed, W. 1955. Social control in the newsroom: A functional analysis. *Social Forces* 33: 326–335.

Brill, A. 1999. Online newspaper advertising: A study of format and integration with news content. In *Advertising and the World Wide Web*, ed. D. Schumann and E. Thorson. Erlbaum.

Bromley, R., and Bowles, D. 1995. Impact of Internet on use of traditional news media. *Newspaper Research Journal* 16, no. 2: 14–27.

Bromley, M., and Tumber, H. 1997. From Fleet street to Cyberspace: The British "Popular" Press in the Late Twentieth Century. *Communications* 22: 365–376.

Brown, N., and Atwater, T. 1986. Videotex news: A content analysis of three videotex services and their companion newspapers. *Journalism Quarterly* 63: 554–561.

Burt, R. 1992. *Structural Holes.* Harvard University Press.

Callon, M. 1986a. Some elements of a sociology of translation: Domestication of the scallops and the fishermen of St. Brieux Bay. In *Power, Action and Belief*, ed. J. Law. Routledge and Kegan Paul.

Callon, M. 1986b. The sociology of an actor-network: The case of the electric vehicle. In *Mapping the Dynamics of Science and Technology*, ed. M. Callon et al. Macmillan.

Cameron, G., Curtin, P., Hollander, B., Nowak, G., and Schamp, S. 1996. Electronic newspapers: Toward a research agenda. *Journal of Mediated Communication* 11, no. 1: 3–53.

Campbell, J., and Thomas, H. 1981. The videotex marketplace—A theory of evolution. *Telecommunications Policy* 5: 111–120.

Campbell-Kelly, M., and Aspray, W. 1996. *Computer: A History of the Information Machine.* Basic Books.

Carey, J. 1986. The dark continent of American journalism. In *Reading the News*, ed. R. Manoff and M. Schudson. Pantheon.

Carey, J., and Pavlik, J. 1993. Videotex: The sword in the stone. In *Demystifying Media Technology*, ed. J. Carey and E. Dennis. Mayfield.

Carlile, P. 2002. A pragmatic view of knowledge and boundaries: Boundary objects in new product development. *Organization Science* 13: 442–455.

Carlson, D. 2000. David Carlson's Online Timeline. Available at http://iml.jou.ufl.edu.

Carlson, W. B. 1992. Artifacts and frames of meaning: Thomas A. Edison, his managers, and the cultural construction of motion pictures. In *Shaping Technology/Building Society*, ed. W. Bijker and J. Law. MIT Press.

Carveth, R., Owers, J., and Alexander, A. 1998. The economics of online media. In *Media Economics*, ed. A. Alexander et al., second edition. Erlbaum.

Case, D. 1994. The social shaping of videotex: How information services for the public have evolved. *Journal of the American Society for Information Science* 45: 483–497.

Case, T. 1994. Fax, CD-ROM, on-line services and newspapers. *Editor & Publisher*, July 16: 12–13, 51.

Castells, M. 2001. *The Internet Galaxy: Reflections on the Internet, Business, and Society*. Oxford University Press.

Ceruzzi, P. 1998. *A History of Modern Computing*. MIT Press.

Chandler, A., Jr. 2001. *Inventing the electronic century: The epic story of the consumer electronics and computer industries*. Free Press.

Chandler, A., Jr., and Cortada, J. 2000. The information age: Continuities and differences. In *A Nation Transformed by Information*, ed. A. Chandler Jr. and J. Cortada. Oxford University Press.

Charity, A. 1995. *Doing Public Journalism*. Guilford.

Charon, J.-M. 1987. Videotex: From interaction to communication. *Media, Culture & Society* 9: 301–332.

Chilton, W. 1982 Newspapers: An endangered species. *Editor & Publisher*, November 20: 31, 44

Chyi, H. I., and Sylvie, G. 1998. Competing with whom? Where? And how? A Structural analysis of the electronic newspaper market. *Journal of Media Economics* 11, no. 2: 1–18.

Clark, A. 1997. *Being There: Putting Brain, Body, and World Together Again*. MIT Press.

Clayman, S., and Reisner, A. 1998. Gatekeeping in action: Editorial conferences and assessments of newsworthiness. *American Sociological Review* 63: 178–199.

Cockburn, C., and Ormrod, S. 1993. *Gender and Technology in the Making*. Sage.

Cohen, J. 1996. Commercial online and newspapers. *Editor & Publisher*, February 17: 40–41.

Cohen, J. 1997. Thomson acquires online classified unit. *Editor & Publisher*, January 25: 24.

Cole, M. 1996. *Cultural Psychology: A Once and Future Discipline*. Belknap.

Collins, H. 1975. The seven sexes: A study in the sociology of a phenomenon, or the replication of experiments in physics. *Sociology* 9: 205–224.

Collins, H. 1981. Stages in the empirical programme of relativism. *Social Studies of Science* 11: 3–10.

Collins, H. 1992. *Changing Order: Replication and Induction in Scientific Practice*, second edition. University of Chicago Press.

Compaine, B. 1980. *The Newspaper Industry in the 1980s: An Assessment of Economics and Technology*. Knowledge Industry Publications.

Compaine, B. 1984. Videotex and the newspaper industry: Threat or opportunity? In *Understanding New Media*, ed. B. Compaine. Ballinger.

Compaine, B. 2000a. The newspaper industry. In *Who Owns the Media*, ed. B. Compaine and D. Gomery. Erlbaum.

Compaine, B. 2000b. The online information industry. In *Who Owns the Media*, ed. B. Compaine and D. Gomery. Erlbaum.

Conniff, M. 1992. Videotext is a terminal case. *Editor & Publisher*, December 19: 23.

Conniff, M. 1995. Mercury Center cites web opportunity. *Editor & Publisher*, February 18: 3.

Coriat, B. 1997. Globalization, variety, and mass production: The metamorphosis of mass production in the new competitive age. In *Contemporary Capitalism*, ed. J. Hollingsworth and R. Boyer. Cambridge University Press.

Cottle, S. 1999. From BBC Newsroom to BBC News Centre: On changing technology and journalist practices. *Convergence* 5, no. 3: 22–43.

Criner, K. 1980. Teletext and videotex in North America: US videotex activities and policy concerns. *Telecommunications Policy* 4: 3–8.

Cummings, J., Sproull, L., and Kiesler, S. 2002. Beyond hearing: Where real-world and online support meet. *Group Dynamics: Theory, Research, and Practice* 6: 78–88.

Darnton, R. 1975. Writing news and telling stories. *Daedalus* 104, spring: 175–194.

Davenport, L. 1987. A Coorientation Analysis of Newspaper Editors' and Readers' Attitudes towards Videotex, Online News and Databases: A Study of Perception and Options. Doctoral dissertation, Ohio University.

David, P. 1986. Understanding the necessity of QWERTY: The necessity of history. In *Economic History and the Modern Economist*, ed. W. Parker. Blackwell.

Desbarats, P. 1981. Newspapers and Computers: An Industry in Transition. Research Studies, Royal Commission on Newspapers, Ottawa, volume 8.

DiMaggio, P. 1992. Nadel's paradox revisited: Relational and cultural aspects of organizational structure. In *Networks and Organizations,* ed. N. Nohria and R. Eccles. Harvard Business School Press.

Dotinga, R. 1999. The great pretenders. *Mediainfo.com,* July: 18, 20, 22.

Douglas, S. 1987. *Inventing American Broadcasting, 1899–1922.* Johns Hopkins University Press.

Dozier, D., and Rice, R. 1984. Rival theories of electronic newsreading. In *The New Media,* ed. R. Rice. Jossey-Bass.

DuBick, M. 1978. The organizational structure of newspapers in relation to their metropolitan environments. *Administrative Science Quarterly* 23: 418–433.

Dunlop, C., and Kling, R., ed. 1991. *Computerization and Controversy: Value Conflicts and Social Choices.* Academic Press.

Edwards, P. 1995. From "impact" to social process: Computers in society and culture. In *Handbook of Science and Technology Studies,* ed. S. Jasanoff et al. Sage.

Edwards, R. 1979. *Contested Terrain: The Transformation of the Workplace in the Twentieth Century.* Basic Books.

Eliasoph, N. 1988. Routines and the making of oppositional news. *Critical Studies in Mass Communication* 5: 313–334.

Elton, M., and Carey, J. 1983. Computerizing information: Consumer reactions to teletext. *Journal of Communication* 33: 162–173.

Elzen, B. 1986. Two ultracentrifuges: A Comparative study of the social construction of artifacts. *Social Studies of Science* 16: 621–662.

Emery, E., and Emery, M. 1978. *The Press and America: An Interpretative History of the Mass Media,* fourth edition. Prentice-Hall.

Endres, F. 1985. Daily newspaper utilization of computer databases. *Newspaper Research Journal* 7, fall: 29–35.

Engeström, Y., Miettinen, R., and Punamäki, R.-L., eds. 1999. *Perspectives on Activity Theory.* Cambridge University Press.

Epstein, E. 1973. *News from Nowhere: Television and the News.* Random House.

Eriksen, L. B., and Sørgaard. 1996. Organisational implementation of WWW in Scandinavian newspapers: Tradition based approaches dominate. Unpublished.

Eriksen, L., and Ihlström, C. 1999. In the Path of the Pioneers: Longitudinal Study of Web News Genre. Unpublished.

Esser, F. 1998. Editorial structures and work principles in British and German newspapers. *European Journal of Communication* 13, no. 3: 375–405.

Ettema, J. 1989. Interactive electronic text in the United States: Can videotex ever go home again? In *Media Use in the Information Age*, ed. J. Salvaggio and J. Bryant. Erlbaum.

Ettema, J., and Glasser, T. 1998. *Custodians of Conscience*. Columbia University Press.

Featherly, K. 1998. Shockwave: The jolt newspapers need? *Mediainfo.com* 18, July: 20–22.

Feenberg, A. 1995. *Alternative Modernity: The Technical Turn in Philosophy and Social Theory*. University of California Press.

Fernandez, R., and Gould, R. 1994. A dilemma of state power: Brokerage and influence in the national health policy domain. *American Journal of Sociology* 99: 1455–1491.

Fidler, R. 1997. *Mediamorphosis: Understanding New Media*. Pine Forge Press.

Finn, T. A., and Stewart, C. 1985. From consumer to organizational videotex applications: Will videotex find a home at the office? *Communication Yearbook* 9: 809–826.

Fischer, C. 1992. *America Calling: A Social History of the Telephone to 1940*. University of California Press.

Fishman, M. 1980. *Manufacturing the News*. University of Texas Press.

Fitzgerald, M. 1984. Videotex verdict: Still unclear. *Editor & Publisher*, May 19: 36, 38.

Fitzgerald, M. 1990a. *Chicago Tribune* folds fax paper. *Editor & Publisher*, August 18: 13.

Fitzgerald, M. 1990b. Grappling with audiotex. *Editor & Publisher*, April 28: 16, 65.

Fitzgerald, M. 1994. Six chains form multimedia research firm. *Editor & Publisher*, April 9: 38.

Fitzgerald, M. 1996. A year of turmoil. *Editor & Publisher*, January 6: 9–14.

Fixmer, R. 1997. A letter to our readers: The new CyberTimes front page. New York Times on the Web, August 27. Available at http://www.nytimes.com.

Flichy, P. 1995. *Dynamics of Modern Communication: The Shaping and Impact of New Communication Technologies*. Sage.

Flynn, L. 1998. Online city guides compete in crowded field. New York Times on the Web, September 14. Available at http://www.nytimes.com.

Friedland, L. 1996. Electronic democracy and the new citizenship. *Media, Culture & Society* 18: 185–212.

Fujimura, J. 1987. Constructing "do-able" problems in cancer research: Articulating alignment. *Social Studies of Science* 17: 257–293.

Fujimura, J. 1996. *Crafting Science: A Sociohistory of the Quest for the Genetics of Cancer.* Harvard University Press.

Galison, P. 1997. *Image and Logic: A Material Culture of Microphysics.* University of Chicago Press.

Galloway, D. 1997a. A look at ourselves. HoustonChronicle.com, February 23. Available at http://www.chron.com.

Galloway, D. 1997b. Great suggestions. HoustonChronicle.com, April 27. Available at http://www.chron.com.

Galloway, D. 1997c. No more traveling light. HoustonChronicle.com, February 9. Available at http://www.chron.com.

Galloway, D. 1997d. Victory at sea. HoustonChronicle.com, April 6. Available at http://www.chron.com.

Gans, H. 1980. *Deciding What's News: A Study of CBS Evening News, NBC Nightly News, Newsweek and Time.* Vintage.

Garcia-Murillo, M., and MacInnes, I. 2001. FCC organizational structure and regulatory convergence. *Telecommunications Policy* 25: 431–452.

Garneau, G. 1989a. Audiotex and newspapers. *Editor & Publisher*, April 22: 84, 86, 88, 90.

Garneau, G. 1989b. Ma Bell gets a green light. *Editor & Publisher*, August 5: 32.

Garneau, G. 1990. Telecommunications: The newspaper buzzword. *Editor & Publisher*, January 6: 20, 24.

Garneau, G. 1993. Major newspaper companies join MIT research project. *Editor & Publisher*, May 15: 13.

Garneau, G. 1995. Campus press races online. *Editor & Publisher*, April 22: 72–74.

Garneau, G. 1996. The Web: Next step in interactive agenda. *Editor & Publisher*, February 17: 2i.

Garrison, B. 1997. Online services, Internet in 1995 newsrooms. *Newspaper Research Journal* 18, no. 3–4: 79–93.

Garud, R., and Rappa, M. 1994. A socio-cognitive model of technology evolution: The case of cochlear implants. *Organization Science* 5: 344–362.

Gasser, L. 1986. The integration of computing and routine work. *ACM Transactions on Office Information Systems* 4: 205–225.

Gersh Hernandez, D. 1996. Advice for the future. *Editor & Publisher*, December 28: 9–13, 34.

Gibson, B. 1983. "On the shelf" software developed for videotex. *Editor & Publisher*, May 7: 64.

Gieber, W. 1964. News is what newspapermen make it. In *People, Society, and Mass Communication*, ed. L. Dexter and D. White. Free Press.

Gillett, S., and Vogelsang, I., eds. 1998. *Competition, Regulation, and Convergence: Current Trends in Telecommunications Policy Research*. Erlbaum.

Giobbe, D. 1996. AT&T phases out interchange. *Editor & Publisher*, January 20: 27.

Gitlin, T. 1980. *The Whole World Is Watching: Mass Media in the Making and Unmaking of the New Left*. University of California Press.

Gloede, B. 1980. NAB conference stresses profits in new technology. *Editor & Publisher*, October 4: 12, 30.

Golightly, D. 1996. Virtual Voyager—Concept and Working Plan. Internal Report, Houston Chronicle Interactive, December 4.

Grabher, G. 2000. Spaces of creativity: Heterarchies in the British advertising industry. Paper presented at Heterarchies seminar, Columbia University.

Granovetter, M. 1973. The strength of weak ties. *American Journal of Sociology* 78: 1360–1380.

Greenhouse, L. 2001. Freelancers win in case of work kept in databases. New York Times on the Web, June 26. Available at http://www.nytimes.com.

Greenstein, S., and Khanna, T. 1997. What does industry convergence mean? In *Competing in the Age of Digital Convergence*, ed. D. Yoffie. Harvard Business School Press.

Grunfeld, D. 1996. All the news that's fit to re-print: Writers vs. the Times. *Columbia Journalism Review*, January-February: 10–11.

Gunther, L. 1997. Enough stress already. HoustonChronicle.com, March 23. Available at http://www.chron.com.

Hall, J. 2001. *Online Journalism: A Critical Primer*. Pluto.

Hamilton, S. 1997. Incomplete determinism: A discourse analysis of cybernetic futurology in early cyberculture. *Journal of Communication Inquiry* 22: 177–204.

Hampton, K., and Wellman, B. 1999. Netville on-line and off-line: Observing and surveying in a wired suburb. *American Behavioral Scientist* 43: 475–492.

Hampton, K., and Wellman, B. 2002. The not so global village of Netville. In *The Internet and Everyday Life*, ed. B. Wellman and C. Haythornthwaite. Blackwell.

Hansen, K., Ward, J., Conners, J., and Neuzil, M. 1994. Local breaking news, sources, technology and news routines. *Journalism and Mass Communication Quarterly* 71, no. 3: 561–572.

Hargadon, A., and Sutton, R. 1997. Technology brokering and innovation in a product development firm. *Administrative Science Quarterly* 42: 716–749.

Harper, C. 1998. *And That's the Way It Will Be: News and Information in a Digital World.* New York University Press.

Harrison, B. 1994. *Lean and Mean: The Changing Landscape of Corporate Power in the Age of Flexibility.* Basic Books.

Haythornthwaite, C. 2001. Introduction: The Internet in everyday life. *American Behavioral Scientist* 45: 363–382.

Hearst, A. 1995. Caught up in the web. *Columbia Journalism Review,* May–June: 62–64.

Hecht, J. 1983. Information services search for identity. *High Technology* 3: 58–65.

Heilbrunn, H. 1994. Picking a partner. *Editor & Publisher,* February 12: 36–37.

Henderson, K. 1999. *On Line and on Paper: Visual Representations, Visual Culture, and Computer Graphics in Design Engineering.* MIT Press.

Henderson, R., and Clark, K. 1990. Architectural innovation: The reconfiguration of existing product technologies and the failure of established firms. *Administrative Science Quarterly* 35: 9–30.

Higgins, M. 2000. Divergent messages in a converging world. *The Information Society* 16: 49–63.

Hirschhorn, L. 1984. *Beyond Mechanization.* MIT Press.

Hotaling, B. 1948. Facsimile broadcasting: Problems and possibilities. *Journalism Quarterly* 25: 139–44.

Houston, F. 1996. Going local. CJR Online, November–December. Available at http://www.cjr.org.

Huenergard, C. 1981. Field enterprises to test teletext. *Editor & Publisher,* April 25: 15.

Huenergard, C. 1982. Milwaukee dailies to offer 24 hr. electronic newspaper. *Editor & Publisher,* May 1: 63.

Hughes, T. 1969. Technological momentum in history: Hydrogenation in Germany 1898–1933. *Past & Present* 44: 106–132.

Hughes, T. 1987. The evolution of large technological systems. In *The Social Construction of Technological Systems,* ed. W. Bijker et al. MIT Press.

Hughes, T. 1994. Technological momentum. In *Does Technology Drive History?* ed. M. Smith and L. Marx. MIT Press.

Hutchins, E. 1995. *Cognition in the Wild.* MIT Press.

Huxford, J. 2000. Cultures in collision: Newspapers and the Internet. Paper presented at annual meeting of International Communication Association, Acapulco.

Ittner, C., and Kogut, B. 1995. How control systems can support organization flexibility. In *Redesigning the Firm*, ed. E. Bowman and B. Kogut. Oxford University Press.

Iwaasa, R. 1988. Convivial messaging systems: Startling facts and figures about electronic mail (messageries) for French households. *The Information Society* 5: 265–269.

Jackson, M., and Paul, N. 1998. *Newspaper Publishing and the World Wide Web*. Poynter Institute for Media Studies.

Jacoby, S. 1985. *Employing Bureaucracy: Managers, Unions, and the Transformation of Work in American Industry, 1900–1945*. Columbia University Press.

Jankowski, N., and van Selm, M. 2000. Traditional news media online: An examination of added values. *Communications: The European Journal of Communication Research* 25: 85–101.

Jasanoff, S. Forthcoming. Introduction. In *States of Knowledge*, ed. S. Jasanoff. Routledge.

Jenkins, H. 2001. Convergence? I diverge. *Technology Review*, June: 93.

Johansen, R., Nyhan, M., and Plummer, R. 1980. Teletext and videotex in North America: Issues and insights for the USA. *Telecommunication Policy* 4: 31–41.

Jones, S., ed. 1995. *Cybersociety: Computer-Mediated Communication and Community*. Sage.

Jones, S., ed. 1997. *Virtual Culture: Identity and Communication in Cybersociety*. Sage.

Kaniss, P. 1991. *Making Local News*. University of Chicago Press.

Katz, M., and Shapiro, C. 1985. Network externalities, competition, and compatibility. *American Economic Review* 75: 424–440.

Kawamoto, K. 1998. News and information at the crossroads: Making sense of the new on-line environment in the context of traditional mass communication study. In *The Electronic Grapevine*, ed. D. Borden and K. Harvey. Erlbaum.

Kenney, K., Gorelik, A., and Mwangi, S. 2000. Interactive features of online newspapers. First Monday 5, no. 1. Available at http://www.firstmonday.dk.

King, E. 1998. Redefining relationships: Interactivity between news producers and consumers. *Convergence* 4, no. 4: 26–32.

Kirsner, S. 1997a. Explosive expansion at Tribune news site. *Editor & Publisher*, July 5: 28–29.

Kirsner, S. 1997b. Profits in site? AJR NewsLink, December. Available at http://ajr.newslink.org.

Kirsner, S. 1997c. The battle for beantown: City guides crowd the Boston metro market. *Editor & Publisher*, November: 12, 14, 16–17, 37.

Kline, R. 2000. *Consumers in the Country: Technology and Social Change in Rural America.* Johns Hopkins University Press.

Kline, R., and Pinch, T. 1996. Users as agents of change: The social construction of the automobile in the rural United States. *Technology and Culture* 37: 763–795.

Kling, R. 1994. Reading "all about" computerization: How genre conventions shape nonfiction social analysis. *The Information Society* 10: 147–172.

Kling, R., and Iacono, S. 1988. The mobilization of support for computerization: The role of computerization movements. *Social Problems* 35: 226–243.

Kogut, B., Shan, W., and Walker, W. 1992. The make-or-cooperate decision in the context of an industry network. In *Networks and Organizations*, ed. R. Eccles and N. Nohria. Harvard Business School Press.

Laakaniemi, R. 1981. The computer connection: America's first computer-delivered newspaper. *Newspaper Research Journal, 2* (4), 61–68.

Landow, G. 1997. *Hypertext 2.0: The Convergence of Contemporary Critical Theory and Technology.* Johns Hopkins University Press.

Lane, D., and Mansfield, R. 1996. Strategy under complexity: Fostering generative relationships. *Long Range Planning* 29, no. 2: 215–231.

Lasica, J. 1996. Net gain. AJR NewsLink, November. Available at http://ajr.newslink.org.

Latour, B. 1987. *Science in Action: How to Follow Scientists and Engineers through Society.* Harvard University Press.

Latour, B. 1993. *We Have Never Been Modern.* Harvard University Press.

Latour, B. 1994. On technical mediation: Philosophy, sociology, genealogy. *Common Knowledge* 3: 29–64.

Lave, J., and Wenger, E. 1992. *Situated Learning: Legitimate Peripheral Participation.* Cambridge University Press.

Law, J. 1987. Technology and heterogeneous engineering: The case of Portuguese expansion. In *The Social Construction of Technological Systems*, ed. W. Bijker et al. MIT Press.

Law, J., and Callon, M. 1988. Engineering and sociology in a military aircraft project: A network analysis of technological change. *Social Problems* 35: 284–297.

Lee, A., and So, C. 2000. Electronic newspaper as digital marketplaces. Paper presented at annual conference of International Communication Association, Acapulco.

LeGates, J. 1984. Changes in the information industries: Strategic implications for newspapers. In *Understanding New Media*, ed. B. Compaine. Ballinger.

Lester, M. 1974. News as Practical Accomplishment: A Conceptual and Empirical Analysis of Newswork. Doctoral dissertation, University of California, Santa Barbara.

Levins, H. 1997a. Attitude adjustment. *Editor & Publisher*, June 28: 44–45.

Levins, H. 1997b. In search of: Internet busine$$. *Editor & Publisher*, February 8: 4i–6i.

Levins, H. 1997c. Largest newspaper web series ever? *Editor & Publisher*, November 29: 22–23.

Levins, H. 1997d. Time of change and challenge. *Editor & Publisher*, January 4: 58–60.

Levins, H. 1998a. Chicago Tribune wins 3 top honors in world's best online newspaper awards. E&P Online, February 6. Available at http://www.mediainfo.com.

Levins, H. 1998b. Connections adapts with changing industry. *Editor & Publisher*, June 27: 20, 22.

Levins, H. 1998c. New York Times joins Classified Ventures. *Editor & Publisher*, August 8: 26.

Li, X. 1998. Web page design and graphic use of three U.S. newspapers. *Journalism and Mass Communication Quarterly* 75, no. 2: 353–365.

Liebeskind, K. 1997. The battle for help wanted. *Editor & Publisher*, February 8: 8i, 10i, 12i, 13i.

Liebeskind, K. 1999. AdOne acquired by five newspaper companies. *Editor & Publisher*, January 30: 34.

Lievrouw, L., Bucy, E., Finn, T. A., Frindte, W., Gershon, R., Haythornthwaite, C., Köhler, T., Metz, J. M., and Sundar, S. S. 2001. Bridging the subdisciplines: An overview of communication and technology research. *Communication Yearbook* 24: 272–296.

Lievrouw, L., and Livingstone, S. 2002. Introduction: The social shaping and consequences of ICTs. In *Handbook of New Media*, ed. L. Lievrouw and S. Livingstone. Sage.

Light, A. 1999. Fourteen Users in Search of a Newspaper: The Effect of Expectation on Online Behaviour. Technical report CSRP 507, University of Sussex.

Light, A., and Rogers, Y. 1999. Conversation as publishing: The role of news forums on the Web. *Journal of Computer-Mediated Communication* 4, no. 4. Available at http://www.ascusc.org.

Lin, C., and Jeffries, L. 1998. Factors influencing the adoption of multimedia cable technology. *Journalism & Mass Communication Quarterly* 75: 341–352.

Mackay, H., and Gillespie, G. 1992. Extending the social shaping of technology approach: Ideology and appropriation. *Social Studies of Science* 22: 685–716.

Mackay, H., Carne, C., Beynon-Davies, P., and Tudhope, D. 2000. Reconfiguring the user: Using Rapid Application Development. *Social Studies of Science* 30: 737–757.

MacKenzie, D. 1984. Marx and the machine. *Technology and Culture* 25: 473–502.

MacKenzie, D., and Wajcman, J. 1985a. Introductory essay: The social shaping of technology. In *The Social Shaping of Technology*, ed, D. MacKenzie and J. Wajcman. Open University Press.

MacKenzie, D., and Wajcman, J., eds. 1985b. *The Social Shaping of Technology*. Open University Press.

Maguire, T. 1982. The diversity principle is now the law. *Editor & Publisher*, August 28: 44.

Maher, M. 1994. Electronic gatekeeper for news. *Editor & Publisher*, June 25: 70, 72, 108.

Manovich, L. 2001. *The Language of New Media*. MIT Press.

Mantooth, S. S. 1982. The Electronic Newspaper: Its Prospects and Directions for Future Study. Doctoral dissertation, University of Tennessee.

Marchand, M. 1987. *La grande aventure du Minitel*. Larousse.

Markus, M. L. 1987. Toward a "critical mass" theory of interactive media: Universal access, interdependence and diffusion. *Communication Research* 14: 491–511.

Marsden, P. 1982. Brokerage behavior in restricted exchange networks. In *Social Structure and Network Analysis*, ed. P. Marsden and N. Lin. Sage.

Martin, S., and Hansen, K. 1998. *Newspapers of Record in a Digital Age: From Hot Type to Hot Link*. Praeger.

Marvin, C. 1980. Delivering the news of the future. *Journal of Communication* 30: 10–20.

Marvin, C. 1988. *When Old Technologies Were New: Thinking about Electric Communication in the Late Nineteenth Century*. Oxford University Press.

Massey, B., and Levy, M. 1999. Interactivity, online journalism, and English-language web newspapers in Asia. *Journalism & Mass Communication Quarterly* 76: 138–151.

Matei, S., and Ball-Rokeach, S. 2001. Real and virtual social ties: Connections in the everyday lives of seven ethnic neighborhoods. *American Behavioral Scientist* 45: 550–564.

Mayntz, R., and Schneider, V. 1988. The dynamics of system development in a comparative perspective: Interactive videotex in Germany, France and Britain. In *The Development of Large Technical Systems*, ed. R. Mayntz and T. Hughes. Westview.

McGrath, J., and Kelly, J. 1986. *Time and Human Interaction: Toward a Social Psychology of Time*. Guilford.

McIntyre, C. 1983. Teletext in the United Kingdom. In *The Future of Videotex*, ed. E. Sigel. Knowledge Industry Publications.

McKenney, J. 1995. *Waves of Change: Business Evolution through Information Technology*. Harvard Business School Press.

McManus, J. 1994. *Market-Driven Journalism: Let the Citizen Beware?* Sage.

McMillan, S. 1998. Who pays for content? Funding in interactive media. *Journal of Computer-Mediated Communication* 4, no. 1. Available at http://www.ascusc.org.

McNair, B. 1998. *The Sociology of Journalism*. Arnold.

Mecca, R. 1981. Newspapers and home video information systems: The present, the promise and the peril. *Videodisc/Teletext* 1, no. 1: 18–29.

Miles, I. 1992. When mediation is the message: How suppliers envisage new markets. In *Contexts of Computer-Mediated Communication*, ed. M. Lea. Harvester Wheatsheaf.

Miles, I., and Thomas, G. 1997. User resistance to new interactive media: Participants, processes and paradigms. In *Resistance to New Technology:*, ed. M. Bauer. Cambridge University Press.

Millard, A. 1995. *America on Record: A History of Recorded Sound*. Cambridge University Press.

Miller, D., and Slater, D. 2000. *The Internet: An Ethnographic Approach*. Berg.

Miller, M. 1997a. "I told you so." HoustonChronicle.com, May. Available at http://www.chron.com.

Miller, M. 1997b. The mail. HoustonChronicle.com, October 26. Available at http://www.chron.com.

Miller, M. 1997c. The mail. HoustonChronicle.com, June 28. Available at http://www.chron.com.

Miller, M. 1998. Cruisers in cyberspace. HoustonChronicle.com, January 21. Available at http://www.chron.com.

Miller, T. 1985a. Newspaper firms involved in electronic publishing. *Editor & Publisher*, March 9: 28, 35.

Miller, T. 1985b. Videotex market shaken up by PC boom. *Editor & Publisher*, March 9. 26–27.

Miller, T. 1988. Databases. *Editor & Publisher*, September 10: 34–35, 38, 40–44.

Misa, T. 1988. How machines make history, and how historians (and others) help them to do so. *Science, Technology, & Human Values* 13: 308–311.

Misa, T. 1992. Controversy and closure in technological change: Constructing "steel." In *Shaping Technology/Building Society*, ed. W. Bijker and J. Law. MIT Press.

Molina, A. 1997a. Issues and challenges in the evolution of multimedia: The case of the newspaper. *Futures* 29: 193–212.

Molina, A. 1997b. Newspapers: The slow walk to multimedia. *Long Range Planning* 30: 218–226.

Molina, A. 1999. Transforming visionary products into realities: Constituency-building and observacting in NewsPad. *Futures* 31: 291–332.

Morris, M., and Ogan, C. 1996. The Internet as mass medium. *Journal of Communication* 46: 39–50.

Mosco, V. 1982. *Pushbutton Fantasies: Critical Perspectives on Videotex and Information Technology*. Ablex.

Moses, L. 1999a. Newspapers, present and future: Good in 1999! Better in 2000? *Editor & Publisher*, December 11: 16.

Moses, L. 1999b. What difference a year makes. *Editor & Publisher*, December 2: 30.

Moses, L. 2000. Heard on the street: Newspapers still a good place to be. *Editor & Publisher*, January 3: 15.

Mott, F. 1962. *American Journalism: A History: 1690–1960*, third edition. Macmillan.

Murray, J. 1999. *Hamlet on the Holodeck: The Future of Narrative in Cyberspace*. MIT Press.

Negroponte, N. 1996. *Being Digital*. Vintage.

Neuberger, C., Tonnemacher, J., Biebl, M., and Duck, A. 1998. Online—the future of newspapers? Germany's dailies on the World Wide Web. *Journal of Computer-Mediated Communication* 4, no. 1. Available at http://www.ascusc.org.

Neustadt, R. 1982. *The Birth of Electronic Publishing: Legal and Economic Issues in Telephone, Cable and Over-the-Air Teletext and Videotex*. Knowledge Industries.

Neuwirth, R. 1998a. Old news makes new business sense. *Editor & Publisher*, April 25: 14, 16–17.

Neuwirth, R. 1998b. Race into cyberspace gushes $80M red ink. *Editor & Publisher*, December 19: 12–13.

Newhagen, J., Cordes, J., and Levy, M. 1995. Nightly@nbc.com: Audience scope and the perception of interactivity in viewer mail on the Internet. *Journal of Communication* 45: 164–175.

Newhagen, J., and Levy, M. 1998. The future of journalism in a distributed communication architecture. In *The Electronic Grapevine*, ed. D. Borden and K. Harvey. Erlbaum.

Newspaper Association of America. 1998. Facts about Newspapers 1998. Available at http://www.naa.org.

Newspaper Association of America. 2001. Facts about newspapers 2001. Available at http://www.naa.org.

New York Times Company. 1997. The New York Times Company Annual Report.

Noack, D. 1997. Publishers 1 freelancers 0. *Editor & Publisher*, August 23: 7–8.

Noack, D. 1999a. Finally, a club for us. *Editor & Publisher*, May 15: 34.

Noack, D. 1999b. Poll says Web news use is mainstream. *Editor & Publisher*, January 16: 26.

Noble, D. 1984. *Forces of Production*. Knopf.

Noll, A. M. 1980. Teletext and videotex in North America: Service and system implications. *Telecommunications Policy* 4: 25–31.

Noll, A. 1985. Videotex: Anatomy of a failure. *Information and Management* 9: 99–109.

O'Donnell, J. 1998. *Avatars of the Word: From Papyrus to Cyberspace*. Harvard University Press.

Orlikowski, W. 2000. Using technology and constituting structures: A practice lens for studying technology in organizations. *Organization Science* 11: 404–428.

Orlikowski, W., and Gash, D. 1994. Technological frames: Making sense of information technology in organizations. *ACM Transactions on Information Systems* 12, no. 2: 174–207.

Outing, S. 1996. Hold on (line) tight. *Editor & Publisher*, February 17: 4i–6i.

Outing, S. 1998a. Community publishing: Coming soon . . . really. E&P Online, May 6. Available at http://www.editorandpublisher.com.

Outing, S. 1998b. Trends to do (online) business by. E&P Online, August 14. Available at http://www.editorandpublisher.com.

Outing, S. 1999a. A status report on online original content, part 2. E&P Online, January 22. Available at http:www.editorandpublisher.com.

Outing, S. 1999b. News site audiences closing in. *Editor & Publisher*, April 3: 29.

Overduin, H. 1986. News judgment and the community connection in the technological limbo of videotex. *Communication* 9: 229–246.

Owen, B. 1999. *The Internet Challenge to Television*. Harvard University Press.

Padgett, J., and Ansell, C. 1993. Robust action and the rise of the Medici, 1400–1434. *American Journal of Sociology* 98: 1259–1319.

Palmer, J., and Eriksen, L. 1999. Digital newspapers explore marketing on the Internet. *Communications of the ACM* 42, no. 9: 32–40.

Park, R. 1925. The natural history of the newspaper. *American Journal of Sociology* 29: 273–289.

Patrick, A., Black, A., and Whalen, T. 1996. CBC radio on the Internet: An experiment in convergence. *Canadian Journal of Communication* 21: 125–140.

Pavlik, J. 1998. *New Media Technology: Cultural and Commercial Perspectives*, second edition. Allyn and Bacon.

Pavlik, J. 2000. The impact of technology on journalism. *Journalism Studies* 1: 229–237.

Pavlik, J. 2001. *Journalism and New Media*. Columbia University Press.

Peng, F., Them, N., and Xiaoming, H. 1999. Trends in online newspapers: A look at the US Web. *Newspaper Research Journal* 20, no. 2: 52–63.

Pfaffenberger, B. 1989. The social meaning of the personal computer: Or, why the personal computer revolution was no revolution. *Anthropological Quarterly* 61: 39–47.

Picard, R. 1998. The economics of the daily newspaper industry. In *Media Economics*, ed. A. Alexander et al., second edition. Erlbaum.

Picard, R., and Brody, J. 1997. *The Newspaper Publishing Industry*. Allyn and Bacon.

Pickering, A. 1995. *The Mangle of Practice: Time, Agency, and Science*. University of Chicago Press.

Pinch, T. 1996. The social construction of technology: A review. In *Technological change*, ed. R. Fox. Harwood.

Pinch, T. 2001. Why you go to a piano store to buy a synthesizer: Path dependence and the social construction of technology. In *Path Dependence and Creation*, ed. R. Garud and P. Karnoe. Erlbaum.

Pinch, T., and Bijker, W. 1984. The social construction of facts and artefacts: or how the sociology of science and the sociology of technology might benefit each other. *Social Studies of Science* 14: 399–441.

Pinch, T., and Trocco, F. 2002. *Analog Days*. Harvard University Press.

Pine, B. J., II. 1993. *Mass Customization: The New Frontier in Business Competition.* Harvard Business School Press.

Piore, M., and Sabel, C. 1984. *The Second Industrial Divide: Possibilities for Prosperity.* Basic Books.

Podolny, J., and Page, K. 1998. Networks form of organization. *Annual Review of Sociology* 24: 57–76.

Pogash, C. 1996. Cyberspace journalism. AJR NewsLink, June. Available at http://ajr.newslink.org.

Pool, I. de Sola. 1983. *Technologies of Freedom.* Belknap.

Poster, M. 1995. *The Second Media Age.* Polity Press.

Poster, M. 2001. *What's the Matter with the Internet?* University of Minnesota Press.

Powell, W. 1990. Neither market nor hierarchy: Network forms of organization. *Research in Organizational Behavior* 12: 295–336.

Powell, W. 1996. Inter-organizational collaboration in the biotechnology industry. *Journal of Institutional and Theoretical Economics* 152: 197–215.

Preston, P., and Kerr, A. 2001. Digital media, nation-states and local cultures: The case of multimedia "content" production. *Media, Culture & Society* 23: 109–131.

Pride, R. 1998. Media critics and newsgroup-embedded newspapers: Making attentive citizens attentive. In *The Public Voice in a Democracy at Risk,* ed. M. Salvador and P. Sias. Praeger.

Radolf, A. 1980a. Knight-Ridder to test home electronic info system. *Editor & Publisher,* April 12: 7–8.

Radolf, A. 1980b. KRN gives sharp picture of Viewtron. *Editor & Publisher,* December 6: 23.

Radolf, A. 1982. New videotex association seeks conciliatory role. *Editor & Publisher,* August 14: 19.

Radolf, A. 1989. The fax revolution. *Editor & Publisher,* April 22: 126.

Rafaeli, S., and LaRose, R. 1993. Electronic bulletin boards and "public goods" explanations of collaborative mass media. *Communication Research* 20: 277–297.

Reardon, J. 2001. The Human Genome Diversity Project: A case study in coproduction. *Social Studies of Science* 31: 357–388.

Reich, L. 1985. *The Making of American Industrial Research: Science and Business at GE and Bell, 1876–1926.* Cambridge University Press.

Rheingold, H. 1994. *The Virtual Community: Homesteading on the Electronic Frontier.* Harper.

Rice, R., and Paisley, W. 1982. The green thumb videotex experiment. *Telecommunications Policy* 6: 223–235.

Rice, R., and Rogers, E. 1980. Reinvention in the innovation process. *Knowledge* 1: 499–514.

Riley, P., Keough, C., Christiansen, T., Meilich, O., and Pierson, J. 1998. Community or colony: The case of online newspapers and the Web. *Journal of Computer-Mediated Communication* 4, no. 1. Available at http://www.ascusc.org.

Roscoe, T. 1999. The construction of the World Wide Web audience. *Media, Culture & Society* 21: 673–684.

Rosen, P. 1993. The social construction of mountain bikes: Technology and posmodernity in the cycle industry. *Social Studies of Science* 23: 479–513.

Rosenberg, J. 1990. Second fax paper competes in Twin Cities. *Editor & Publisher*, March 17: 31, 59.

Rosenberg, J. 1992. Knight-Ridder info design lab. *Editor & Publisher*, October 31: 26–29, 33.

Rosenberg, J. 1994a. IFRA creates INES. *Editor & Publisher*, February 12: 38–40.

Rosenberg, J. 1994b. Times Mirror dailies on line. *Editor & Publisher*, November 12: 34–35.

Rosenberg, J. 1994c. *Washington Post* to go on line with Ziff-Davis. *Editor & Publisher*, April 9: 40–41.

Rosenberg, J. 1995. Newspaper technology review. *Editor & Publisher*, January 7: 56–64.

Rosenkopf, L., and Tushman, M. 1994. The coevolution of technology and organization. In *Evolutionary Dynamics of Organizations*, ed. J. Baum and J. Singh. Oxford University Press.

Roshco, B. 1975. *Newsmaking*. University of Chicago Press.

Sabel, C. 1991. Moebius-strip organizations and open labor markets: Some consequences of the reintegration of conception and execution in a volatile economy. In *Social Theory for a Changing Society*, ed. P. Bourdieu and J. Coleman. Westview.

Sabel, C., and Zeitlin, J. 1997. Stories, strategies, structures: Rethinking historical alternatives to mass production. In *Worlds of Possibilities*, ed. C. Sabel and J. Zeitlin. Cambridge University Press.

Salomon, G., ed. 1993. *Distributed Cognitions: Psychological and Educational Considerations*. Cambridge University Press.

Saussure, F. de. 1908–09. *Course in General Linguistics*. (Cited here: 1983 Duckworth edition.)

Savell, L. 1996. The Internet and the law. *Editor & Publisher*, September 28: 22–23, 36.

Schiller, D. 1999. *Digital Capitalism: Networking the Global Market System.* MIT Press.

Schlesinger, P. 1978. *Putting "Reality" Together: BBC News.* Methuen.

Schmidt, K., and Bannon, L. 1992. Taking CSCW seriously: Supporting articulation work. *Computer Supported Cooperative Work* 1: 7–40.

Schneider, V., Charon, J., Miles, I., Thomas, G., and Vedel, T. 1991. The dynamics of videotex development in Britain, France and Germany: A cross-national comparison. *European Journal of Communication* 6: 187–212.

Schudson, M. 1978. *Discovering the News.* Basic Books.

Schudson, M. 1982. The politics of narrative form: The emergence of news conventions in print and television. *Daedalus* 111: 97–113.

Schudson, M. 1995. *The Power of News.* Harvard University Press.

Schudson, M. 1997. The sociology of news production revisited. In *Mass Media and Society*, ed. J. Curran and M. Gurevitch, second edition. Arnold.

Schultz, T. 1999. Interactive options in online journalism: A content analysis of 100 U.S. newspapers. *Journal of Computer-Mediated Communication* 5, no. 1. Available at http://www.ascusc.org.

Schultz, T. 2000. Mass media and the concept of interactivity: An exploratory study of online forums and reader email. *Media, Culture & Society* 22: 205–221.

Scofield, J. 1984. The mission of newspaper libraries. *Editor & Publisher*, February 18: 41, 52.

Scranton, P. 1994. Determinism and indeterminacy in the history of technology. In *Does Technology Drive History?* ed. M. Smith and L. Marx. MIT Press.

Seib, P. 2001. *Going Live: Getting the News Right in a Real-Time, Online World.* Rowman and Littlefield.

Shaiken, H. 1984. *Work Transformed.* Lexington Books.

Shefrin, D. 1949. The Radio Newspaper and Facsimile Broadcasting. Master's thesis, University of Missouri.

Shoemaker, P. 1991. *Gatekeeping.* Sage.

Sigal, L. 1973. *Reporters and Officials: The Organization and Politics of Newsmaking.* Heath.

Sigel, E. 1980. Videotext in the U.S. In *Videotext*, ed. E. Sigel. Knowledge Industry Publications.

Sigel, E. 1983a. Introduction. In *The Future of Videotex*, ed. E. Sigel. Knowledge Industry Publications.

Sigel, E. 1983b. Videotext in other countries. In *The Future of Videotex*, ed. E. Sigel. Knowledge Industry Publications.

Sigelman, L. 1973. Reporting the news: An organizational analysis. *American Journal of Sociology* 79: 132–151.

Silverstein, J. 1983. Videotext in the United States. In *The Future of Videotex*, ed. E. Sigel. Knowledge Industry Publications.

Simon, B. 1999. Undead science: Making sense of cold fusion after the (arti)fact. *Social Studies of Science* 29: 61–85.

Singer, J. 1998. Online journalists: Foundations for research into their changing roles. *Journal of Computer-Mediated Communication* 4, no. 1. Available at http://www.ascusc.org.

Singer, J., Tharp, M., and Haruta, A. 1999. Online staffers: Superstars or second-class citizens? *Newspaper Research Journal* 20, no. 3: 29–47.

Slevin, J. 2000. *The Internet and Society*. Polity.

Smith, A. 1979. *The Newspaper: An International History*. Thames and Hudson.

Smith, A. 1980. *Goodbye Gutenberg: The Newspaper Revolution of the 1980s*. Oxford University Press.

Smith, M. 1991. New age journalism. *Editor & Publisher*, January 26: 2TC-3TC, 12TC, 14TC, 18TC-20TC, 22TC-24TC.

Smith, M. 1994. Technological determinism in American Culture. In *Does Technology Drive History?* ed. M. Smith and L. Marx (pp. 1–35. MIT Press.

Smith, M., and Marx, L., eds. 1994. *Does Technology Drive History? The Dilemma of Technological Determinism*. MIT Press.

Smulyan, S. 1994. *Selling Radio: The Commercialization of American Broadcasting, 1920–1934*. Smithsonian Institution Press.

Sokolski, J. 1989. News reporting and professionalism: Some constraints on the reporting of the news. *Media, Culture and Society* 11: 207–228.

Sommer, P. 1983. Videotext in the United Kingdom. In *The Future of Videotex*, ed. E. Sigel. Knowledge Industry Publications.

Sproull, L. 2000. Computers in the U.S. households since 1977. In *A Nation Transformed by Information*, ed. A. Chandler Jr. and J. Cortada. Oxford University Press.

Sproull, L., and Kiesler, S. 1991. *Connections: New Ways of Working in the Networked Organization*. MIT Press.

Star, S. L. 1991. The sociology of the invisible: The primacy of work in the writings of Anselm Strauss. In *Social Organization and Social Process*, ed. D. Maines. Aldine de Gruyter.

Star, S. L., and Griesemer, J. 1989. Institutional ecology, translations, and boundary objects: Amateurs and professionals in Berkeley's Museum of Vertebrate Zoology, 1907–39. *Social Studies of Science* 19: 387–420.

Stark, D. 1996. Recombinant property in East European capitalism. *American Journal of Sociology* 101: 993–1027.

Stark, D. 2001. Ambiguous assets for uncertain environments: Heterarchy in post-socialist firms. In *The 21st Century Firm*, ed. P. DiMaggio. Princeton University Press.

Stark, R. 1962. Policy and the pros: An organizational analysis of a metropolitan newspaper. *Berkeley Journal of Sociology* 7: 11–31.

Staudenmaier, J. 1989. *Technology's Storytellers: Reweaving the Human Fabric.* MIT Press.

Stein, M. L. 1993. First step to a multimedia future. *Editor & Publisher*, April 10: 18–19.

Stepp, C. 1996. The new journalist. AJR NewsLink, April. Available at http://ajr.newslink.org.

Stone, G. 1987. *Examining Newspapers: What Research Reveals about America's Newspapers.* Sage.

Stone, M. 1998. Editorial and ads benefit from new technology. *Editor & Publisher*, October 17: 32.

Stone, M. 1999. Desperately seeking geeks and hucksters for Web ventures. *Editor & Publisher*, January 16: 30–32.

Strauss, A. 1985. Work and the division of labor. *Sociological Quarterly* 26: 1–19.

Strauss, A. 1988. The articulation of a project work: An organizational process. *Sociological Quarterly* 29: 163–178.

Streeck, W. 1991. On the institutional conditions of diversified quality production. In *Beyond Keynesianism*, ed. E. Matzner and W. Streeck. Elgar.

Suchman, L. 1996. Supporting articulation work. In *Computerization and Controversy*, ed. R. Kling, second edition. Academic Press.

Suchman, L. 2000. Working relations of technology production and use. Paper presented at Heterarchies Seminar, Columbia University.

Sullivan, C. 1999. The Times they are a-targeting. *Editor & Publisher*, April 17: 44–45.

Sumpter, R. 2000. Daily newspaper editors' audience construction routines: A case study. *Critical Studies in Media Communication* 17: 334–346.

Sunstein, C. 2001. *republic.com.* Princeton University Press.

Sylvie, G., and Witherspoon, P. 2002. *Time, Change, and the American Newspaper.* Erlbaum.

Tankard, J., and Ban, H. 1998. Online newspapers: Living up to their potential? Paper presented at annual convention of Association for Journalism and Mass Communication, Baltimore.

Teubner, G. 1993. The many-headed hydra: Networks as higher-order collective actors. In *Corporate Control and Accountability*, ed. J. McCahery et al. Clarendon.

Thalhimer, M. 1994. High tech news or just "shovelware"? *Media Studies Journal,* winter: 41–52.

Thompson, J. 1995. *The Media and Modernity: A Social Theory of the Media.* Polity.

Thorson, E., Wells, W., and Rogers, S. 1999. Web advertising's birth and early childhood as viewed in the pages of *Advertising Age.* In *Advertising and the World Wide Web*, ed. D. Schumann and E. Thorson. Erlbaum.

Tuchman, G. 1973. Making news by doing work: Routinizing the unexpected. *American Journal of Sociology* 79: 110–131.

Tuchman, G. 1978. *Making News: A Study in the Construction of Reality.* Free Press.

Turkle, S. 1995. *Life on the Screen: Identity in the Age of Internet.* Simon and Schuster.

Turow, J. 1997. *Breaking Up America: Advertisers and the New Media World.* University of Chicago Press.

Tushman, M., and Murmann, J. 1998. Dominant designs, technology cycles, and organizational outcomes. *Research in Organizational Behavior* 20: 231–266.

Tushman, M., and Rosenkopf, L. 1992. Organizational determinants of technological change: Toward of sociology of technological evolution. In *Research in Organizational Behavior*, ed. B. Staw and L. Cummings. JAI.

Tydeman, J., Lipinski, H., Adler, R., Nyhan, M., and Zwimpfer, L. 1982. *Teletext and Videotex in the United States: Market Potential, Technology and Public Policy Issues.* McGraw-Hill.

Tyler, M. 1979. Videotex, Prestel and teletext: The economics and politics of some electronic publishing media. *Telecommunications Policy* 3: 37–51.

Tyre, M., and Orlikowski, W. 1994. Windows of opportunity: Temporal patterns of technological adaptations in organizations. *Organization Science* 5: 98–118.

van Cuilenburg, J., and Verhoest, P. 1998. Free and equal access: In search of policy models for converging communication systems. *Telecommunications Policy* 22: 171–181.

Vedel, T., and Charon, J. M. 1989. Videotex in France: The invention of a mass-medium? In Pathways to Telematics, ed. V. Schneider et al. (unpublished).

Warren, P. 1967. The Metropolitan Newspaper as a Political Institution: An Organizational Analysis of the New York Press. Doctoral dissertation, Harvard University.

Wayner, P. 1996. Computer simulations: New-Media tools for online journalism. New York Times on the Web, October 9. Available at http://www.nytimes.com.

Weaver, D. 1983. Videotex Journalism: Teletext, Viewdata, and the News. Erlbaum.

Weaver, D., and Wilhoit, G. C. 1986. The American Journalist: A Portrait of U.S. News People and Their Work. Indiana University Press.

Webb, W. 1995a. Flat panel newspaper fells flat. Editor & Publisher, August 12: 32.

Webb, W. 1995b. Interactive classified service launched. Editor & Publisher, October 28: 38.

Webb, W. 1995c. Washington Post debuts its Digital Ink online service. Editor & Publisher, July 29: 25, 30.

Wertsch, J. 1998. Mind as Action. Oxford University Press.

Westney, D. E. 1987. Imitation and Innovation: The Transfer of Western Organizational Patterns to Meiji Japan. Harvard University Press.

White, D. 1949. The "gatekeeper": A case study in the selection of news. Journalism Quarterly 27: 383–390.

Wilkinson, M. 1980. Viewdata: The Prestel system. In Videotext, ed. E. Sigel. Knowledge Industry Publications.

Williams, R. 1974. Television: Technology and Cultural Form. Wesleyan University Press.

Williams, W. 1998. The blurring of the line between advertising and journalism in the on-line environment. In The Electronic Grapevine, ed. D. Borden and K. Harvey. Erlbaum.

Williams, R., and Edge, D. 1996. The social shaping of technology. Research Policy 25: 865–899.

Williamson, O. 1991. Comparative economic organization: The analysis of discrete structural alternatives. Administrative Science Quarterly 36: 269–296.

Willis, J. 1994. The Age of Multimedia and Turbonews. Praeger.

Winner, L. 1986. Mythinformation. In The Whale and the Reactor, ed. L. Winner. University of Chicago Press.

Winston, B. 1998. Media Technology and Society: A History from the Telegraph to the Internet. Routledge.

Woolgar, S. 1991. Configuring the user: The case of usability trials. In *A Sociology of Monsters*, ed. J. Law. Routledge.

Wynne, B. 1988. Unruly technology: Practical rules, impractical discourses and public understanding. *Social Studies of Scienc* 18: 147–167.

Yates, J. 1989. *Control through Communication: The Rise of System in American Management.* Johns Hopkins University Press.

Yates, J. 1993. Co-evolution of information-processing technology and use: Interaction between the life insurance and tabulating industries. *Business History Review* 67: 1–53.

Zavoina, S., and Reichert, T. 2000. Media convergence/management change: The evolving workflow for visual journalists. *Journal of Media Economics* 13: 1143–1151.

Zollman, P. 1998. New media hiring: Vets offer tips. *Editor & Publisher*, March 21: 20–21, 41.

Zuboff, S. 1988. *In the Age of the Smart Machine.* Basic Books.

Inside Technology: The Series

Donald MacKenzie, *Inventing Accuracy: A Historical Sociology of Nuclear Missile Guidance*

Donald MacKenzie, *Knowing Machines: Essays on Technical Change*

Donald MacKenzie, *Mechanizing Proof: Computing, Risk, and Trust*

Maggie Mort, *Building the Trident Network: A Study of the Enrollment of People, Knowledge, and Machines*

Nelly Oudshoorn and Trevor Pinch, editors, *How Users Matter: The Co-Construction of Users and Technology*

Paul Rosen, *Framing Production: Technology, Culture, and Change in the British Bicycle Industry*

Susanne K. Schmidt and Raymund Werle, *Coordinating Technology: Studies in the International Standardization of Telecommunications*

Dominque Vinck, editor, *Everyday Engineering: An Ethnography of Design and Innovation*

Index

DATE DUE

GAYLORD

PRINTED IN U.S.A